# SpringerBriefs in Energy

T0281487

More information about this series at http://www.springer.com/series/8903

Shaharin Anwar Sulaiman
Ramani Kannan · Samsul Ariffin Abdul Karim
Nursyarizal Mohd Nor
Editors

# Sustainable Electrical Power Resources through Energy Optimization and Future Engineering

 Springer

*Editors*
Shaharin Anwar Sulaiman
Department of Mechanical Engineering
Universiti Teknologi PETRONAS (UTP)
Seri Iskandar, Perak
Malaysia

Ramani Kannan
Department of Electrical
   and Electronic Engineering
Universiti Teknologi PETRONAS (UTP)
Seri Iskandar, Perak
Malaysia

Samsul Ariffin Abdul Karim
Department of Fundamental
   and Applied Sciences
Universiti Teknologi PETRONAS (UTP)
Seri Iskandar, Perak
Malaysia

Nursyarizal Mohd Nor
Department of Electrical
   and Electronic Engineering
Universiti Teknologi PETRONAS (UTP)
Seri Iskandar, Perak
Malaysia

ISSN 2191-5520          ISSN 2191-5539  (electronic)
SpringerBriefs in Energy
ISBN 978-981-13-0434-7      ISBN 978-981-13-0435-4   (eBook)
https://doi.org/10.1007/978-981-13-0435-4

Library of Congress Control Number: 2018941210

Printed on acid-free paper

This Springer imprint is published by the registered company Springer Nature Singapore Pte Ltd. part of Springer Nature
The registered company address is: 152 Beach Road, #21-01/04 Gateway East, Singapore 189721, Singapore

# Preface

The global demand for electrical power in countries has been increasing quite steadily, except for a 1.5% cut-down in 2009, i.e. the first time since World War II. The growth in Asia has increased significantly in recent years, especially in China. Energy efficiency and environment-friendly features are also critical issues in modern electricity production and transmission, and thus, intensive research activities within these areas are imperative. Motivated by such issues, this book shares the efforts by researchers in science and engineering on conventional and renewable energy, energy efficiency, and optimization.

The editors would like to express their gratitude to all the contributing authors for their effort and dedication in preparing the manuscripts for the book. It is hoped that the book will serve as an important reference to interested readers.

Seri Iskandar, Malaysia

Shaharin Anwar Sulaiman
Ramani Kannan
Samsul Ariffin Abdul Karim
Nursyarizal Mohd Nor

# Contents

# Contributors

**Muhammad Adlan Abdul Jalil** Department of Electrical and Electronic Engineering, Universiti Teknologi PETRONAS (UTP), Seri Iskandar, Perak, Malaysia

**Samsul Ariffin Abdul Karim** Fundamental and Applied Sciences Department and Centre for Smart Grid Energy Research (CSMER), Institute of Autonomous System, Universiti Teknologi PETRONAS (UTP), Seri Iskandar, Perak, Malaysia

**Mohd Faris Abdullah** Department of Electrical and Electronic Engineering, Universiti Teknologi PETRONAS (UTP), Seri Iskandar, Perak, Malaysia

**Asmala Ahmad** Faculty of Information Technology and Communication, Universiti Teknikal Malaysia Melaka, Durian Tunggal, Melaka, Malaysia

**Moustafa Ahmed** Universiti Teknologi PETRONAS (UTP), Seri Iskandar, Perak, Malaysia

**Abid Ali** Department of Electrical and Electronic Engineering, Universiti Teknologi PETRONAS (UTP), Seri lskandar, Perak, Malaysia

**Aravind CV** School of Engineering, Taylor's University, Subang Jaya, Malaysia

**Zuhairi Baharuddin** Department of Electrical and Electronic Engineering, Universiti Teknologi PETRONAS (UTP), Seri Iskandar, Perak, Malaysia

**Siti Rahimah Batcha** Fundamental and Applied Sciences Department, Universiti Teknologi PETRONAS (UTP), Seri Iskandar, Perak, Malaysia

**Deva Brindha** Electrical and Electronics Engineering, Eswari Engineering College, Chennai, India

**I. Daniel** Electrical and Electronics Engineering, Taylor's University, Subang Jaya, Malaysia

**Khairul Nisak Md Hasan** Universiti Teknologi PETRONAS (UTP), Seri Iskandar, Perak, Malaysia

**Ser Yi Heng** School of Engineering, Taylor's University, Subang Jaya, Malaysia

**Taib Ibrahim** Department of Electrical and Electronic Engineering, Universiti Teknologi PETRONAS (UTP), Seri lskandar, Perak, Malaysia

**Ramani Kannan** Department of Electrical and Electronic Engineering, Universiti Teknologi PETRONAS (UTP), Seri lskandar, Perak, Malaysia

**Sujatha Krishnan** Research Associate with the Centre for Future Learning, Taylor's University, Subang Jaya, Malaysia

**Nursyarizal Mohd Nor** Department of Electrical and Electronic Engineering, Universiti Teknologi PETRONAS (UTP), Seri lskandar, Perak, Malaysia

**Mahmod Othman** Fundamental and Applied Sciences Department, Universiti Teknologi PETRONAS (UTP), Seri Iskandar, Perak, Malaysia

**Mohd Fakhizan Romlie** Department of Electrical and Electronic Engineering, Universiti Teknologi PETRONAS (UTP), Seri lskandar, Perak, Malaysia

**Hamzah Sakidin** Fundamental and Applied Sciences Department, Universiti Teknologi PETRONAS (UTP), Seri Iskandar, Perak, Malaysia

**Alex Suresh** Electrical and Electronic Engineering, S.A. Engineering College, Chennai, India

# Chapter 1
# Load Frequency Control in the Deregulated Environment for the Future Green Energy Network

**Aravind CV, Ser Yi Heng, Ramani Kannan, Deva Brindha and Sujatha Krishnan**

With the increasing number, micro generation grids are being emerged into the power system network, whereas the future of the power system network is highly promising. Hence, the increase of load demand and fluctuation of load effect the frequency of the system which is one of the major elements in power quality. Load Frequency Control (LFC) is a method to maintain the frequency in the power system when fluctuation of load occurs. Controller is used to perform Load Frequency Control in the system which collects the data (error) and generates a control signal to the system. This chapter focused on a research regarding the Load Frequency Control method in power system with the addition of a renewable unit such as a speed varying standalone wind turbine. The study included simulation of two-area power system in deregulated environment by using MATLAB/Simulink. The output of the power system such as settling time, overshooting and under-shooting are studied.

Aravind CV (✉) · S. Y. Heng
School of Engineering, Taylor's University, Subang Jaya, Malaysia
e-mail: aravind_147@yahoo.com

R. Kannan
Department of Electrical and Electronic Engineering, Universiti Teknologi PETRONAS, Perak Darul Ridzuan, Malaysia

D. Brindha
Department of Electrical and Electronic Engineering, Eswari Engineering College, Chennai, India

S. Krishnan
Research Associate with the Centre for Future Learning, Taylor's University, Subang Jaya, Malaysia

## 1.1 Introduction

The population in a developing country such as Malaysia grows rapidly which leads to increase the demand for power supply. The power quality becomes a major issue in the design of power system to meet the load demand. The structure of an interconnected power system in a deregulated environment is more complex than the conventional power system. In the deregulated environment, generation and distribution belong to different companies, so the consumer is bonded to Distribution Company who also further bonded with generating company from any area. The fluctuation of load demand brings an issue to the power system to maintain the power supply quality such as fixed voltage and frequency. Besides that, more power generation is needed to handle the rise of load demand. To achieve the sustainability of the power supply, a stable power system is required to supply fixed frequency and voltage [1–6].

Currently, the researchers and scientists studied and developed a smart control to the power system based on the load demand side to maintain the power supply quality [7–12]. Due to the rise in load demand, more power generation is required to connect with the existing power system. The smart control of the system is important to maintain the grid of the system in order to lower the stress to the existing generation and energy losses. The proposed design of the power system in deregulated environment with a different controller, which minimizes the settling of the frequency. However, few limitations such as external disturbance in the modelling are ignored and the number of DISCO and GENCO are limited to four.

## 1.2 Power System Model

In the traditional power system, the supplier generated the power, transmitted and distributed it to the consumer. In the deregulated power system or restructured power system, the generation, transmission and distribution belongs to Generation Company (GENCO), Transmission Company (TRANSCO) and Distribution Company (DISCO). Figures 1.1 and 1.2 show the sketch of a traditional power system and a deregulated power system. In the traditional power system, one company owns the whole power system from generation to consumer. The consumer only gets the power from this company. In the deregulated environment, one or more GENCO and DISCO is installed in one area. The GENCO and DISCO collaborate to contract with any type of company in the interconnected area. In return, the consumer can contract with any of the DISCO from the same area [13, 14].

**Fig. 1.1** Conventional power system

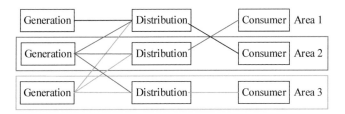

**Fig. 1.2** Power system in deregulated environment

In the deregulated power system, the particular DISCO can establish a contract with any GENCO. The DISCO has no contract with the GENCO in the same area. The DISCO Participation Matrix (DPM) was introduced to understand the contract between DISCO and GENCO in any area easier, which is given by:

$$DPM = \begin{bmatrix} cpf_{ij} & \cdots & cpf_{ij} \\ \vdots & \ddots & \vdots \\ cpf_{ij} & \cdots & cpf_{ij} \end{bmatrix}, \quad DPM = \begin{bmatrix} cpf_{11} & cpf_{12} & cpf_{13} & cpf_{14} \\ cpf_{21} & cpf_{22} & cpf_{23} & cpf_{24} \\ cpf_{31} & cpf_{32} & cpf_{33} & cpf_{34} \\ cpf_{41} & cpf_{42} & cpf_{43} & cpf_{44} \end{bmatrix} \quad (1.1)$$

Figure 1.3 shows a sketch of deregulated two-area power system. In Eq. (1.1), the row of the matrix represented as GENCO and the column represented as DISCO. The element in the matrix represented the fraction of total load contracted by the distribution company $j$ from the generation company $i$, $cpf$ is contract participation factor. The summation of all elements in one column is equal to one. The scheduled steady state tie line power flow and actual tie line power are represented by:

$$\Delta P_{tie12,sch} = (Demand\ of\ DISCOs\ in\ area\ 1\ to\ GENCOs\ in\ area2) \\ - (Demand\ of\ DISCOs\ in\ area\ 2\ to\ GENCOs\ in\ area\ 1) \quad (1.2)$$

$$\Delta P_{tie12,actual} = \frac{2\pi T_{12}}{s}(\Delta F_1 - \Delta F_2) \quad (1.3)$$

The tie line power error given by:

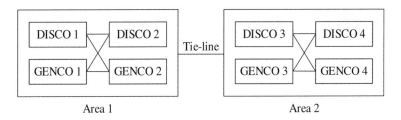

**Fig. 1.3** Deregulated two-area power system

$$\Delta P_{tie12,error} = \Delta P_{tie12,actual} - \Delta P_{tie12,sch} \tag{1.4}$$

became zero when power flow on the tie line achieved scheduled power flow. The controller collects the input signal Area Control Error (ACE) to produce the required action control signals, as given by:

$$ACE_1 = B_1 \Delta F_1 + \Delta P_{tie12,error}, \quad ACE_2 = B_2 \Delta F_2 + a_{12} \Delta P_{tie12,error} \tag{1.5}$$

## 1.3  Two Area Power System Model

Figure 1.4 shows the conventional two-area power system model. In any instance if the load is changed, the power delivered from Area 1 given by:

$$P_{tie,1} = \frac{|V_1||V_2|}{X_{12}} \sin\left(\delta_1^o - \delta_2^o\right) \tag{1.6}$$

where subscripts 1 and 2 represent Area 1 and Area 2, respectively. The power angle of the two area is represented by $\left(\delta_1^o - \delta_2^o\right)$. For changes in $\delta_1$ and $\delta_2$, the tie-line power is expressed by:

$$\Delta P_{tie,1}(pu) = T_{12}(\Delta\delta_1 - \Delta\delta_2) \tag{1.7}$$

where

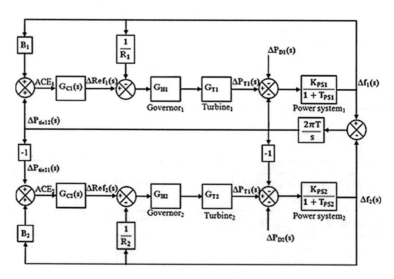

**Fig. 1.4**  Conventional two-area power system

$$T_{12} = \frac{|V_1||V_2|}{P_{r1}X_{12}}\cos(\delta_1^o - \delta_2^o) = Synchronizing\ Coefficient \qquad (1.8)$$

The change of the power output in tie-line Area 1 is given by:

$$\Delta P_{tie,1} = 2\pi T_{12}\left(\int \Delta f_1 dt - \int \Delta f_2 dt\right) \qquad (1.9)$$

where the change of the power output in tie-line Area 2 is given by:

$$\Delta P_{tie,2} = 2\pi T_{21}\left(\int \Delta f_2 dt - \int \Delta f_2 dt\right) \qquad (1.10)$$

$$T_{21} = \frac{|V_2||V_1|}{P_{r2}X_{21}}\cos(\delta_2^o - \delta_1^o) = Synchronizing\ Coefficient \qquad (1.11)$$

where, $\Delta f_1$ and $\Delta f_2$: change of frequency in Area 1 and Area 2.

From Eqs. (1.9) and (1.10), the steady state operational capability is indicated as sensitive to the change in frequency. So, Eqs. (1.9) and (1.10) must be equal to zero, which means the system has no change in frequency. The incremental power balance equation for Area 1 and Area 2 are given as:

$$\Delta P_{G1} - \Delta P_{D1} = \frac{2H_1}{f_1^0}\frac{d}{dt}(\Delta f_1) + B_1\Delta f_1 + \Delta P_{tie,1} \qquad (1.12)$$

$$\Delta P_{G2} - \Delta P_{D2} = \frac{2H_2}{f_2^0}\frac{d}{dt}(\Delta f_2) + B_2\Delta f_2 + \Delta P_{tie,2} \qquad (1.13)$$

Taking Laplace transform of Eqs. (1.12) and (1.13) results in:

$$\Delta F_1(s) = [\Delta P_{G1}(s) - \Delta P_{D1}(s) - \Delta P_{tie,1}(s)] \times \frac{K_{ps1}}{1 + T_{ps1}s} \qquad (1.14)$$

$$\Delta F_2(s) = [\Delta P_{G2}(s) - \Delta P_{D2}(s) - \Delta P_{tie,2}(s)] \times \frac{K_{ps2}}{1 + T_{ps2}s} \qquad (1.15)$$

Taking Laplace Transform of Eqs. (1.9) and (1.10), $\Delta P_{tie,1}(s)$ and $\Delta P_{tie,2}(s)$ results in:

$$\Delta P_{tie,1}(s) = \frac{2\pi T_{12}}{s}[\Delta F_1(s) - \Delta F_2(s)] \qquad (1.16)$$

$$\Delta P_{tie,2}(s) = \frac{2\pi T_{21}}{s}[\Delta F_2(s) - \Delta F_1(s)] \qquad (1.17)$$

The Area Controller Error (ACE) in Area 1 and Area 2 can be represented by:

$$ACE_1 = B_1 \Delta F_1 + \Delta P_{tie12,error}, \quad ACE_2 = B_2 \Delta F_2 + a_{12} \Delta P_{tie12,error} \quad (1.18)$$

### 1.3.1 Deregulated Two-Areas Power System

The tie-line power between Area 1 and Area 2 in a deregulated power system is given by (Fig. 1.5):

$$\begin{aligned}
\Delta P_{tie12,sch} &= \sum_{i=1}^{2} \sum_{j=3}^{4} cpf_{ij} \Delta P_{Lj} - \sum_{i=3}^{4} \sum_{j=1}^{2} cpf_{ij} \Delta P_{Lj} \\
&= (cpf_{13} + cpf_{23}) \Delta P_{L3} + (cpf_{14} + cpf_{24}) \Delta P_{L4} \\
&\quad - (cpf_{31} + cpf_{41}) \Delta P_{L1} - (cpf_{32} + cpf_{42}) \Delta P_{L2}
\end{aligned} \quad (1.19)$$

ACE Participation Factor (APF) is the component that distributes to each participating GENCO. The total of the APF in each area must be equal to unity as:

$$APF_1 + APF_2 = 1, \quad APF_3 + APF_4 = 1 \quad (1.20)$$

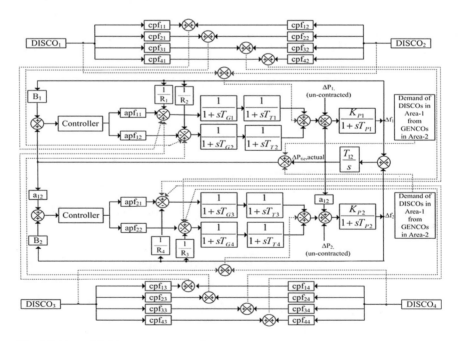

**Fig. 1.5** Deregulated two-area power system

### 1.3.2   Proposed Structure of Deregulated Two-Area Power System

The variation frequency of new design of power system can be written as:

$$\Delta F_1(s) = \left[\Delta P_{G1}(s) - \Delta P_{D1}(s) - \Delta P_{tie,1}(s)\right]$$
$$\times \left(\frac{K_{ps1}}{1 + T_{ps1}s} + \frac{K_{ps,new}}{1 + T_{ps,new}s}\right) \qquad (1.21)$$

$$\Delta F_2(s) = \left[\Delta P_{G2}(s) - \Delta P_{D2}(s) - \Delta P_{tie,2}(s)\right]$$
$$\times \left(\frac{K_{ps1}}{1 + T_{ps1}s} + \frac{K_{ps,new}}{1 + T_{ps,new}s}\right) \qquad (1.22)$$

Table 1.1 shows the corresponding value for the parameters used in the modeling design.

## 1.4   Findings

### 1.4.1   Case 1

In this case, assume all the GENCOs in each area are equal as shown in Table 1.2. The sum of GENCOs in each area must be equal to 1. Assume the load fluctuation

**Table 1.1** Parameter of power system

| Parameter | Value |
|---|---|
| $T_{G1}, T_{G2}, T_{G3}, T_{G4}$ | 0.08 |
| $T_{R1}, T_{R2}, T_{R3}, T_{R4}$ | 5 |
| $T_{T1}, T_{T2}, T_{T3}, T_{T4}$ | 10 |
| $T_{P1}, T_{P2}$ | 20 |
| $K_{P1}, K_{P2}$ | 120 |
| $T_{12}$ | 3.424 |
| $R_1, R_2, R_3, R_4$ | 2.4 |
| $B_1, B_2$ | 0.4256 |
| $a_{12}$ | −1 |

**Table 1.2** GENCOs in Each Area

| Parameter | Value |
|---|---|
| GENCO 1 in area 1, $apf_1$ | 0.5 |
| GENCO 2 in area 1, $apf_2$ | $apf_2 = 1 - apf_1 = 0.5$ |
| GENCO 3 in area 2, $apf_3$ | 0.5 |
| GENCO 4 in area 2, $apf_4$ | $apf_4 = 1 - apf_3 = 0.5$ |

happens only in Area 1, the value of each contract participation factor that (cpf) as in Eq. (1.23), with the load demand of all DISCOs equal to 0.1. In the DPM, it shows that $DISCO_3$ and $DISCO_4$ do no need power from any GENCO where cpf in Area 2 are equal to 0. The demand of $DISCO_1$ and $DISCO_2$ identical to the GENCO in Area 1.

$$DPM = \begin{bmatrix} cpf_{11} & cpf_{12} & cpf_{13} & cpf_{14} \\ cpf_{21} & cpf_{22} & cpf_{23} & cpf_{24} \\ cpf_{31} & cpf_{32} & cpf_{33} & cpf_{34} \\ cpf_{41} & cpf_{42} & cpf_{43} & cpf_{44} \end{bmatrix} = \begin{bmatrix} 0.5 & 0.5 & 0 & 0 \\ 0.5 & 0.5 & 0 & 0 \\ 0 & 0 & 0 & 0 \\ 0 & 0 & 0 & 0 \end{bmatrix} \quad (1.23)$$

### 1.4.2   Case 2

In this case, assume all the GENCOs participated in the Load Frequency Control and each *apf* value is shows in Table 1.3. Besides that, the load demand of all DISCOs are equal to 0.1. The value of each contract participation factor (cpf) is shows in DPM by:

$$DPM = \begin{bmatrix} cpf_{11} & cpf_{12} & cpf_{13} & cpf_{14} \\ cpf_{21} & cpf_{22} & cpf_{23} & cpf_{24} \\ cpf_{31} & cpf_{32} & cpf_{33} & cpf_{34} \\ cpf_{41} & cpf_{42} & cpf_{43} & cpf_{44} \end{bmatrix} = \begin{bmatrix} 0.5 & 0.25 & 0 & 0.3 \\ 0.2 & 0.25 & 0 & 0 \\ 0 & 0.25 & 1 & 0.7 \\ 0.3 & 0.25 & 0 & 0 \end{bmatrix} \quad (1.24)$$

The DISCO in Area 1 and Area 2 has established a contract with the GENCO in their own area and GENCO from other area.

### 1.4.3   Case 3

In this case, assume one of the DISCO needs more power than the amount of agreement earlier had in the contract with the GENCO. So, the power is to be provided in the same area and this becomes the load in that area. With the modification of case 2, $DISCO_1$ needs excess of 0.1 power. The new value for each DISCO is as in Table 1.4. The GENCO and DPM are same as the environment in case 2.

**Table 1.3** *apf* of each GENCOs

| Parameter | Value |
|---|---|
| GENCO 1 in area 1, $apf_1$ | 0.75 |
| GENCO 2 in area 1, $apf_2$ | $apf_2 = 1 - apf_1 = 0.25$ |
| GENCO 3 in area 2, $apf_3$ | 0.5 |
| GENCO 4 in area 2, $apf_4$ | $apf_4 = 1 - apf_3 = 0.5$ |

**Table 1.4**  DISCO in each area

| Parameter | Value |
|---|---|
| DISCO 1 in area 1, $DISCO_1$ | 0.1 + 0.1 = 0.2 |
| DISCO 2 in area 1, $DISCO_2$ | 0.1 |
| DISCO 3 in area 2, $DISCO_3$ | 0.1 |
| DISCO 4 in area 2, $DISCO_4$ | 0.1 |

**Table 1.5**  Proposed Design Power System

| | | PI | | | ANFIS | | |
|---|---|---|---|---|---|---|---|
| | | Settling time (s) | Peak overshoot | Peak undershoot | Settling time (s) | Peak overshoot | Peak undershoot |
| Case-1 | Area 1 | 168.59 | 1.86 | −5.272 | 14.59 | 1.39 | −0.47 |
| | Area 2 | 168.48 | 1.694 | −5.261 | 14.49 | 1.28 | −0.46 |
| | Tie-line | 120.16 | 1.562 | 0 | 3.97 | −0.32 | 0 |
| Case-2 | Area 1 | 129.16 | 2.276 | 0 | 12.46 | 1.64 | 0 |
| | Area 2 | 129.05 | 2.361 | 0 | 12.36 | 1.75 | 0 |
| | Tie-line | 80.779 | 0 | −0.7438 | 7.70 | 0 | −0.54 |
| Case-3 | Area 1 | 129.12 | 2.324 | 0 | 12.60 | 1.68 | −0.17 |
| | Area 2 | 129 | 2.409 | 0 | 12.50 | 1.78 | −0.17 |
| | Tie-line | 80.708 | 0 | −0.743 | 7.71 | 0.10 | −0.55 |

The new design of power system design in a deregulated environment is created to maintain the power quality of the system such as frequency. The proposed design has better performance compared to conventional design in terms of three parameters such as settling time, overshooting and undershooting. The proposed design has 36, 43 and 44% improvement on settling time, overshooting and undershooting as compared to the conventional design. Although there are few oscillations generated, but the fast settling time makes this research more valuable and meaningful. The summary of findings is shown in Table 1.5.

The results show that the ANFIS and FLC has a similar result but much better than a PI controller in terms of settling, undershooting and overshooting. The settling is reduced from 168.59 to 14.554. The undershooting is reduced from −5.272 to −0.4688. The overshooting is reduced from 1.86 to 1.387.

## 1.5   Summary

A new Load Frequency Control (LFC) method for two-areas in power system using MATLAB/SIMULINK is modelled. Based on the given results, there is a increment more than 20% in the settling time, 40% in the undershoot value by increasing 30% overshooting. The proposed system shows 91.26% improvement on settling time, also 91.06% in undershoot improvement by increasing 25.38% overshoot. Yet, there are a few limitations, including the surge in the number of areas and different types of generation that increase the complexity in the controller design.

## References

1. S.K. Pandey, S.R. Mohanty, N. Kishor, A literature survey on load-frequency control for conventional and distribution generation power systems. Renew. Sustain. Energy Rev. **25**, 318–334 (2013)
2. A.P.S. Meliopoulos, G.J. Cokkinides, A.G. Bakirtzis, Load-frequency control service in a deregulated environment. Proc. thirty-first Hawaii international conference on systems sciences, vol. 3, no. C, pp. 1–8, 1998
3. S. Srikanth, Load frequency control in deregulated power system using fuzzy C-means. Int. J. Comput. App. **74**(11), 34–41 (2013)
4. W. Tan, Y. Hao, D. Li, Load frequency control in deregulated environments via active disturbance rejection. Int. J. Electr. Power Energy Syst. **66**, 166–177 (2015)
5. L. Dong, Y. Zhang, Z. Gao, A robust decentralized load frequency controller for interconnected power systems. ISA Trans. **51**(3), 410–419 (2012)
6. N. Hasan, S. Farooq, Real time simulation of automatic generation control for interconnected power system. Int. J. Electr. Eng. Inform. **4**(1), 40–51 (2012)
7. K. Das, P. Das, S. Sharma, Load frequency control using classical controller in an isolated single area and two area reheat thermal power system. Int. J. Emerg. Technol. Adv. Eng. **2**(3), 403–409 (2012)
8. S. Subha, Load frequency control with fuzzy logic controller considering governor dead band and generation rate constraint non-linearity. World Appl. Sci. J. **29**(8), 1059–1066 (2014)
9. I. Kocaarslan, Fuzzy logic controller in interconnected electrical power systems for load-frequency control. Int. J. Electr. Power Energy Syst. **27**, 542–549 (2005)
10. K.R. Sudha, R.V. Santhi, Electrical power and energy systems robust decentralized load frequency control of interconnected power system with generation rate constraint using type-2 fuzzy approach. Int. J. Electr. Power Energy Syst. **33**(3), 699–707 (2011)
11. F. Beaufays, Y. Abdel-Magid, B. Widrow, Application of neural networks to load-frequency control in power systems. Neural Netw. **7**(1), 183–194 (1994)
12. M. Pal, K. Kaur, To control load frequency by using integral. Int. J. Innovative Res. Sci. Eng. Technol. **3**(5), 12502–12506 (2014)
13. F.F. Wu, F.L. Zheng, F.S. Wen, Transmission investment and expansion planning in a restructured electricity market. Energy **31**, 954–966 (2006)
14. T.S. Gorripotu, R.K. Sahu, S. Panda, AGC of a multi-area power system under deregulated environment using redox flow batteries and interline power flow controller. Int. J. Eng. Sci. Technol. **18**(4), 555–578 (2015)

# Chapter 2
# Forecasting Solar Radiation Data Using Gaussian and Polynomial Fitting Methods

**Muhammad Adlan Abdul Jalil, Samsul Ariffin Abdul Karim, Zuhairi Baharuddin, Mohd Faris Abdullah and Mahmod Othman**

Solar radiation prediction is an important task prior to installation of solar photovoltaic panel or to estimate the amount solar radiation received at certain time with respective location. The main objective of the work presented in this chapter is to perform fitting for solar radiation data by using polynomials, Gaussian function and sine fitting as well as Jain's methods. Comparison among all methods are measured by using Root Mean Square Error (RMSE) and the coefficient of determination, $R^2$. Two fitting models are proposed for the prediction of solar radiation in the campus of Universiti Teknologi PETRONAS (UTP), Malaysia.

## 2.1  Introduction

Solar radiation data that collected by the solarimeter or solar tracking data system can be used to form a simple fitting model that can represents the series of the solar radiation data. This model can be constructed by using various fitting methods and the best fitting method is used for the solar radiation data prediction or forecast. There exist many researches regarding solar radiation data fitting. For instance,

S. A. Abdul Karim (✉)
Department of Fundamental and Applied Sciences and Centre for Smart Grid Energy Research (CSMER), Institute of Autonomous System, Universiti Teknologi PETRONAS (UTP), Bandar Seri Iskandar, 32610 Seri Iskandar, Perak DR, Malaysia
e-mail: samsul_ariffin@utp.edu.my

M. A. Abdul Jalil · Z. Baharuddin · M. F. Abdullah
Department of Electrical and Electronic Engineering, Universiti Teknologi PETRONAS (UTP), Bandar Seri Iskandar, 32610 Seri Iskandar, Perak DR, Malaysia

M. Othman
Department of Fundamental and Applied Sciences, Universiti Teknologi PETRONAS (UTP), Bandar Seri Iskandar, 32610 Seri Iskandar, Perak DR, Malaysia

© The Author(s), under exclusive license to Springer Nature Singapore Pte Ltd. 2018
S. A. Sulaiman et al. (eds.), *Sustainable Electrical Power Resources through Energy Optimization and Future Engineering*, SpringerBriefs in Energy,
https://doi.org/10.1007/978-981-13-0435-4_2

Al-Sadah et al. [1] and Genc et al. [2] investigated the application of quadratic polynomial fitting for solar radiation data. They concluded that quadratic fitting is the best. Karim and Singh [3] also showed that quadratic polynomial fitting would be suitable to model the series of the solar radiation in Universiti Teknologi PETRONAS (UTP). Besides that, Karim et al. [3–6], Karim [7] and Karim and Kong [8] also discussed fitting methods for solar radiation and temperature data set by using wavelets based approach. Sulaiman et al. [9] analysts the residuals of the daily solar radiation time series. Genc et al. [2] studied the application of cubic spline for solar radiation data received in Izmir, Turkey. Sen [10] and Khatib et al. [11] gave the detail solar radiation modelling. Wang [12] and Hansen et al. [13] discussed various methods for data fitting and data smoothing.

In this study, various fitting methods for solar radiation data are investigated. The numerical comparison between all the methods will be done and the best fitting methods are chosen based on the values of RMSE and the coefficient of determination, $R^2$. Two prediction models are proposed for the solar radiation data prediction.

## 2.2 Methodology

For the work methodology, it is divided into several stages. This include data collection in UTP until construction of the prediction based model for global solar radiation data. Figure 2.1 shows the framework of the research methodology that is used in this study.

**Fig. 2.1** Data fitting flow chart

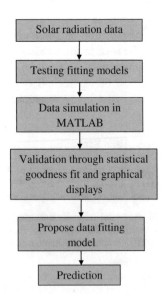

## 2.3   Data Collection

The solar radiation is emitted by the electromagnetic radiation through the sun with wavelength of between 280 and 4000 nm [1, 10]. Furthermore, it was recorded that, the intensity of solar radiation on outside of earth's atmosphere is about 1353 W/m$^2$. The UTP is located at longitude 100.9382 °E and latitude 4.3704 °N. In UTP, the solar radiation data is monitored as well as being measured by the sun tracking system that has equipped with solarimeter. Figure 2.2 shows the solar tracking system. This global solar radiation data is collected daily data for every 30 min. The significant solar radiation data are from 8 am till 6 pm. In this study, the averaged measured solar radiation data sets for January 2011 are used.

## 2.4.   Curve Fitting Methods

There exist various types of data fitting methods or in statistics it is known as data regression [4, 14]. The fitting methods can be either the polynomial based or non-polynomial. In this study, four types of curve fitting methods are considered;

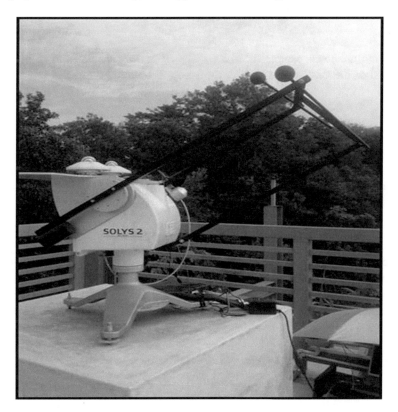

**Fig. 2.2** Sun tracking system equipped with solarimeter

i.e. Gaussian fitting, polynomial fitting (quadratic and cubic), sine fitting as well as Jain's fitting method.

## 2.4.1 Gaussian Function for Curve Fitting

The Gaussian fitting is defined as:

$$\hat{y} = \sum_{j=1}^{M} a_j e^{\left(-((x-b_j)/c_j)^2\right)}.\tag{2.1}$$

where $M$ is the number of terms used and $x$ is the original data. The real coefficients $a_j$, $b_j$ and $c_j$ need to be found. For instance, for Gaussian fitting with one term and two terms, the equations are given as:

$$\hat{y}_{ga1} = a_1 e^{\left(-((x-b_1)/c_1)^2\right)}.\tag{2.2}$$

and

$$\hat{y}_{ga2} = a_1 e^{\left(-((x-b_1)/c_1)^2\right)} + a_2 e^{\left(-((x-b_2)/c_2)^2\right)}.\tag{2.3}$$

where the unknown parameters $a_1, b_1, c_1, a_2, b_2$ and $c_2$ are determined by minimization of the least square error of the fitting method.

## 2.4.2 Polynomial Fitting

The polynomial data fitting or regression model can be described by the following elaborations. For the given observation data $(x_i, y_i)$, $i = 1, 2, \ldots, N$, the regression model (or fitting) is defined as:

$$y_i = f(x_i) + \varepsilon_i, i = 1, 2, \ldots, N.\tag{2.4}$$

where $f$ is a regression (or fitting) function and $\varepsilon_i$ are zero-mean independent random error with a common variance $\sigma^2$. For polynomial fitting, let $f(x) = a_0 + a_1 x + a_2 x^2 + \cdots + a_n x^n$ where $n$ is a positive integer and the degree of the polynomial. Now, say $N > n+1$ then we may fit the data by using least square approach. Let the error of fitting model is given as:

$$e_i = y_i - f(x_i) = y_i - \left\{a_0 + a_1 x_i + a_2 x_i^2 + \cdots + a_n x_i^n\right\}\tag{2.5}$$

Taking sum square of the error given in Eq. (2.5) lead to:

$$S = \sum_{i=0}^{N} e_i^2 = \sum_{i=0}^{N} \left[ y_i - \left\{ a_0 + a_1 x_i + a_2 x_i^2 + \cdots + a_n x_i^n \right\} \right]^2 \tag{2.6}$$

The least square fitting is obtained if the sum of error in Eq. (2.6) is minimized. Hence,

$$\frac{\partial S}{\partial a_i} = 0, \; i = 0, 1, \ldots, n. \tag{2.7}$$

From Eq. (2.7), the following system of linear equations can be obtained:

$$BA = C \tag{2.8}$$

where

$$B = \begin{bmatrix} N & \sum x_i & \sum x_i^2 & \cdots & \sum x_i^n \\ \sum x_i & \sum x_i^2 & \sum x_i^3 & \cdots & \sum x_i^{n+1} \\ \sum x_i^2 & \sum x_i^3 & \sum x_i^4 & \cdots & \sum x_i^{n+2} \\ \vdots & \vdots & \vdots & \ddots & \vdots \\ \sum x_i^n & \sum x_i^{n+1} & \sum x_i^{n+2} & \cdots & \sum x_i^{2n} \end{bmatrix}, A = \begin{Bmatrix} a_0 \\ a_1 \\ a_2 \\ \vdots \\ a_n \end{Bmatrix},$$

and

$$C = \begin{Bmatrix} \sum y_i \\ \sum x_i y_i \\ \sum x_i^2 y_i \\ \vdots \\ \sum x_i^n y_i \end{Bmatrix}.$$

Equation (2.8) can be solved by using Cholesky's method or Gaussian elimination (with pivoting if needed). If $n = 2$, the least square fitting in Eq. (2.4) is called quadratic regression method (or quadratic fitting) given by:

$$y = a_0 + a_1 x + a_2 x^2 \tag{2.9}$$

The system of linear in Eq. (2.5) becomes:

$$\begin{bmatrix} N & \sum x_i & \sum x_i^2 \\ \sum x_i & \sum x_i^2 & \sum x_i^3 \\ \sum x_i^2 & \sum x_i^3 & \sum x_i^4 \end{bmatrix} \begin{Bmatrix} a_0 \\ a_1 \\ a_2 \end{Bmatrix} = \begin{bmatrix} \sum y_i \\ \sum x_i y_i \\ \sum x_i^2 y_i \end{bmatrix}. \tag{2.10}$$

### 2.4.3 Sine Data Fitting

The sine fitting method is defined as:

$$y = \sum_{j=1}^{M} a_j \sin(b_j + c_j) \tag{2.11}$$

The formula for sine fitting with one term or two terms are given in Eqs. (2.12) and (2.13) respectively:

$$y_{s1} = a_1 \sin(b_1 x + c_1) \tag{2.12}$$

$$y_{s2} = a_1 \sin(b_1 x + c_1) + a_2 \sin(b_2 x + c_2) \tag{2.13}$$

where the unknown parameters $a_1, b_1, c_1, a_2, b_2$ and $c_2$ is determined by using least square approach.

### 2.4.4 Jain's Fitting Method

The Jain's method is one of the Gaussian form. The Jain's fitting method is defined as [1, 16]:

$$y = \frac{1}{\sigma\sqrt{2\pi}} \exp\left(-\frac{(h-m)^2}{2\sigma^2}\right) \tag{2.14}$$

where $h$, $m$ and $\sigma$ are determined from the solar radiation data [1, 16].

## 2.5 Error Measurement

There exists many statistical goodness fit measurements. For instance, RMSE, $R^2$ value, Adjusted $R^2$ value, Mean Biased (MB), Mean Square Error (MSE) and Sum Square Error (SSE) etc. In this study we adopt the same error measurement as discussed in Karim and Singh [3] i.e. RMSE and $R^2$ value. RMSE can be calculated by using the following formula:

$$RMSE = \sqrt{\left(\sum_{j=1}^{M}(y_j - \hat{y}_j)^2\right)}. \tag{2.15}$$

where $y_j$ is an original data and $\hat{y}_j$ is fitting data.

## 2.6   Numerical Results

The Gaussian fitting, polynomial fitting (quadratic and cubic), sine fitting and Jain's Fitting methods are used to fit the global solar radiation data collected in UTP as shown in Fig. 2.3. Figure 2.4 shows the polynomial data fitting using quadratic and cubic polynomial fitting. Figures 2.5 and 2.6 show the results for Gaussian and sine fitting methods.

For Jain's method, two difference value of peak value, m is used. When m = 11 and when m = 14. For m = 11, the highest solar radiation value is occurred at 11.00 am meanwhile form = 14, the highest value for solar radiation data is at 2.00 pm. Figures 2.7 and 2.8 show the results.

Figure 2.9 shows the comparison between Jain's method and Gaussian fitting methods. From the figure, we notice that data fitting with Gaussian with one term is better than Jains method. Finally Fig. 2.10 shows the combination for the graphs of all fitting methods.

Based on Figs. 2.4, 2.5, 2.6, 2.7, 2.8, 2.9 and 2.10, Tables 2.1 and 2.2, it can be concluded that polynomial fitting with degree quadratic and Gaussian function is suitable for solar radiation data fitting. Even though fitting data by using polynomial with degree cubic give less RMSE (97.85) and a little bit higher in $R^2$ value (0.8975) as compared to quadratic fitting with RMSE is 105.9 and $R^2$ is 0.8746.

Based on previous works including the studies by Wu and Chan [15], the quadratic polynomial fitting is the best for solar radiation data. Sen [10] also proposed quadratic polynomial fitting to model the solar radiation data at various locations in Turkey. On the other hand, for non-polynomial fitting among three difference methods i.e. Gaussian fitting, sine fitting and Jain's method, it is noted that Gaussian fitting with one term give the best results based with RMSE (60.43)

**Fig. 2.3** Global solar radiation data collected at UTP solar lab

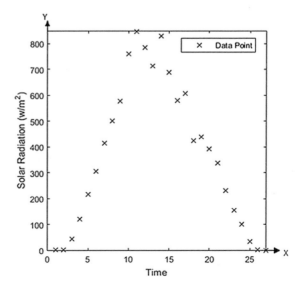

**Fig. 2.4** Results for
polynomial data fitting

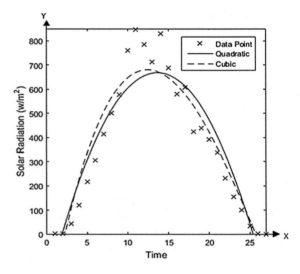

**Fig. 2.5** Results for
Gaussian data fitting

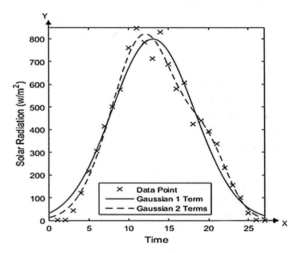

and $R^2$ (0.9592). Besides that, Jain's method seems not suitable for our global solar radiation data. From the graphs in Figs. 2.8 and 2.9, the fitting curve by Jain's method deviate too much from the actual data. This will increase the error and will reduce the $R^2$ value. The value of RMSE values for m = 11 is 105.9 and when m = 14 is 97.85. Table 2.3 shows that Jan's method is not suitable for solar radiation data fitting.

Comparison between quadratic polynomial fitting with Gaussian fitting (one term) indicates that Gaussian fitting is the best with smaller RMSE value and higher $R^2$ value. This result is not discussed in the work of Wu and Chan [15]. Table 2.4 shows the difference between both fitting methods.

**Fig. 2.6** Results for sine data fitting

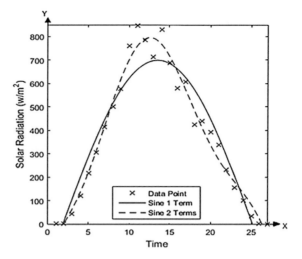

**Fig. 2.7** Jain's method with m = 11

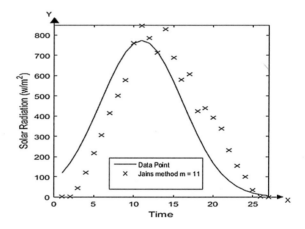

**Fig. 2.8** Jain's method with m = 14

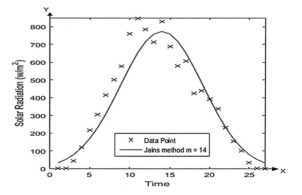

**Fig. 2.9** Comparison of Jain's method and Gaussian data fitting

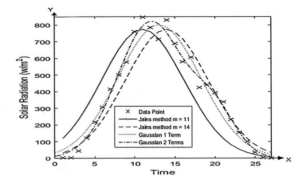

**Fig. 2.10** Overall results for data fitting

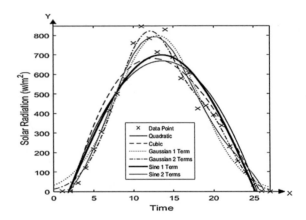

**Table 2.1** RMSE value and $R^2$ for polynomial fitting

| Data fitting method | | Statistical goodness fit | |
|---|---|---|---|
| Polynomial | Degree | RMSE | $R^2$ |
| | 2 | 105.9 | 0.8746 |
| | 3 | 97.85 | 0.8975 |

**Table 2.2** RMSE value and $R^2$ for non-polynomial fitting

| Data fitting method | | Statistical goodness fit | |
|---|---|---|---|
| Type | Term | RMSE | $R^2$ |
| Gaussian | 1 | 60.43 | 0.9592 |
| | 2 | 42.43 | 0.9782 |
| Sine | 1 | 89.26 | 0.9035 |
| | 2 | 50.05 | 0.9755 |

**Table 2.3** Comparison of Jain's method against the actual data

| X Coordinate | Time (Hours) | True value | Jain's method with value | | Differences true value with Jain's method with | |
|---|---|---|---|---|---|---|
| | | | M = 11 | M = 14 | M = 11 | M = 14 |
| 1 | 0630 | 0.0740 | 118.1185 | 32.3014 | −118.0450 | −32.2274 |
| 2 | 0700 | 0.4500 | 168.8004 | 51.6704 | −168.3500 | −51.2204 |
| 3 | 0730 | 42.7380 | 232.3311 | 79.6051 | −189.5930 | −36.8671 |
| 4 | 0800 | 120.2580 | 307.9780 | 118.1185 | −187.7200 | 2.1395 |
| 5 | 0830 | 216.6060 | 393.1971 | 168.8004 | −176.5910 | 47.8056 |
| 6 | 0900 | 305.9940 | 483.4809 | 232.3311 | −177.4870 | 73.6629 |
| 7 | 0930 | 413.6740 | 572.5675 | 307.9780 | −158.8940 | 105.6960 |
| 8 | 1000 | 500.1440 | 653.0590 | 393.1971 | −152.9150 | 106.9469 |
| 9 | 1030 | 577.9400 | 717.3919 | 483.4809 | −139.4520 | 94.4591 |
| 10 | 1100 | 759.3760 | 758.9951 | 572.5675 | 0.3809 | 186.8085 |
| 11 | 1130 | 847.9220 | 773.3922 | 653.0590 | 74.5298 | 194.8630 |
| 12 | 1200 | 785.6440 | 758.9951 | 717.3919 | 26.6489 | 68.2521 |
| 13 | 1230 | 712.3920 | 717.3919 | 758.9951 | −4.9999 | −46.6031 |
| 14 | 1300 | 830.4120 | 653.0590 | 773.3922 | 177.3530 | 57.0198 |
| 15 | 1330 | 687.4600 | 572.5675 | 758.9951 | 114.8925 | −71.5351 |
| 16 | 1400 | 579.8000 | 483.4809 | 717.3919 | 96.3191 | −137.5920 |
| 17 | 1430 | 605.7840 | 393.1971 | 653.0590 | 212.5869 | −47.2750 |
| 18 | 1500 | 424.1780 | 307.9780 | 572.5675 | 116.2000 | −148.3900 |
| 19 | 1530 | 439.7460 | 232.3311 | 483.4809 | 207.4149 | −43.7349 |
| 20 | 1600 | 390.5980 | 168.8004 | 393.1971 | 221.7976 | −2.5991 |
| 21 | 1630 | 337.9320 | 118.1185 | 307.9780 | 219.8135 | 29.9540 |
| 22 | 1700 | 231.5060 | 79.6051 | 232.3311 | 151.9009 | −0.8251 |
| 23 | 1730 | 155.5160 | 51.6704 | 168.8004 | 103.8456 | −13.2844 |
| 24 | 1800 | 100.0700 | 32.3014 | 118.1185 | 67.7686 | −18.0485 |
| 25 | 1830 | 34.4680 | 19.4482 | 79.6051 | 15.0198 | −45.1371 |
| 26 | 1900 | 0.6440 | 11.2776 | 51.6704 | −10.6336 | −51.0264 |
| 27 | 1930 | 0.0000 | 6.2984 | 32.3014 | −6.2984 | −32.3014 |

## 2.7  Solar Radiation Prediction

In this section, the forecasting model for global solar radiation data is proposed. Based on the main results from the previous section, the quadratic polynomial fitting and Gaussian fitting is suitable to fitting the global solar radiation in UTP, Malaysia. Thus it is proposed that two fitting model be used to predict the amount of solar radiation received in UTP for over the years:

**Table 2.4** Comparison of quadratic polynomial and Gaussian (one term) against the actual data

| X coordinate | Time (hours) | True value | Quadratic polynomial value | Gaussian (one term) | Differences with true value | |
|---|---|---|---|---|---|---|
| | | | | | Quadratics | Gaussian |
| 1 | 0630 | 0.0740 | 0.0000 | 51.7800 | 0.0740 | −51.7060 |
| 2 | 0700 | 0.4500 | 13.0800 | 79.9800 | −12.6300 | −79.5300 |
| 3 | 0730 | 42.7380 | 120.9000 | 119.0000 | −78.1620 | −76.2620 |
| 4 | 0800 | 120.2580 | 219.1000 | 170.5000 | −98.8420 | −50.2420 |
| 5 | 0830 | 216.6060 | 307.6000 | 235.2000 | −90.9940 | −18.5940 |
| 6 | 0900 | 305.9940 | 386.4000 | 312.6000 | −80.4060 | −6.6060 |
| 7 | 0930 | 413.6740 | 455.5000 | 400.1000 | −41.8260 | 13.5740 |
| 8 | 1000 | 500.1440 | 514.9000 | 493.2000 | −14.7560 | 6.9440 |
| 9 | 1030 | 577.9400 | 564.7000 | 585.6000 | 13.2400 | −7.6600 |
| 10 | 1100 | 759.3760 | 604.7000 | 669.7000 | 154.6760 | 89.6760 |
| 11 | 1130 | 847.9220 | 635.1000 | 737.5000 | 212.8220 | 110.4220 |
| 12 | 1200 | 785.6440 | 655.7000 | 782.3000 | 129.9440 | 3.3440 |
| 13 | 1230 | 712.3920 | 666.7000 | 799.2000 | 45.6920 | −86.8080 |
| 14 | 1300 | 830.4120 | 668.0000 | 786.3000 | 162.4120 | 44.1120 |
| 15 | 1330 | 687.4600 | 659.6000 | 745.2000 | 27.8600 | −57.7400 |
| 16 | 1400 | 579.8000 | 641.5000 | 680.1000 | −61.7000 | −100.3000 |
| 17 | 1430 | 605.7840 | 613.7000 | 597.8000 | −7.9160 | 7.9840 |
| 18 | 1500 | 424.1780 | 576.3000 | 506.1000 | −152.1220 | −81.9220 |
| 19 | 1530 | 439.7460 | 529.4000 | 413.1000 | −89.6540 | 26.6460 |
| 20 | 1600 | 390.5980 | 472.3000 | 324.1000 | −81.7020 | 66.4980 |
| 21 | 1630 | 337.9320 | 405.8000 | 245.1000 | −67.8680 | 92.8320 |
| 22 | 1700 | 231.5060 | 329.6000 | 178.6000 | −98.0940 | 52.9060 |
| 23 | 1730 | 155.5160 | 243.7000 | 125.3000 | −88.1840 | 30.2160 |
| 24 | 1800 | 100.0700 | 148.1000 | 84.6500 | −48.0300 | 15.4200 |
| 25 | 1830 | 34.4680 | 42.7800 | 55.0900 | −8.3120 | −20.6220 |
| 26 | 1900 | 0.6440 | 0.0000 | 34.5300 | 0.6440 | −33.8860 |
| 27 | 1930 | 0.0000 | 0.0000 | 20.8400 | 0.0000 | −20.8400 |

$$G(x) = a_1 e^{\left(-((x-b_1)/c_1)^2\right)} \tag{2.16}$$

where $a_1 = 799.3, b_1 = 13.07$ and $c_1 = 7.925$ with 95% confidence interval. Similarly, a polynomial quadratic fitting also can be used to predict the solar radiation data as given by:

$$P(x) = a_0 + a_1 x + a_2 x^2 \tag{2.17}$$

where $a_0 = -231.7, a_1 = 132.1$ and $a_2 = -7.844$ with 95% confidence interval.

## 2.8 Summary

In this study, several fitting methods are tested to the global solar radiation data obtained in UTP, Malaysia. From the numerical results as well as graphical displays, quadratic polynomial fitting and Gaussian fitting with one term is suitable for the purpose. Two fitting models are proposed in order to cater the solar radiation prediction. This study has improved the results obtained in Karim and Singh [3], Wu and Chan [15] and Al-Sadah et al. [1]. Furthermore, fitting method for solar radiation data collected for every 30 s in UTP is underway. For future study the new weighted fitting method based on model given in Eqs. (2.16) and (2.17), respectively is proposed.

## References

1. F.H. Al-Sadah, F.M. Ragab, M.K. Arshad, Hourly solar radiation over Bahrain. Energy **15**, 395–402 (1990)
2. A. Genc, I. Kinaci, G. Oturanc, A. Kurnaz, S. Bilir, N. Ozbalta, Statistical analysis of solar radiation data using cubic spline functions. Energy Sources Part A Recovery Utilization and Environmental Effects. **24**, 1131–1138 (2002)
3. S.A.A. Karim, B.S.M. Singh, Global solar radiation modeling using polynomial fitting. Appl. Math. Sci. **8**, 367–378 (2013)
4. S.A.A. Karim, B.S.M. Singh, R. Razali, N. Yahya, B.A. Karim. *Solar Radiation Data Analysis by Using Daubechies Wavelets*. In Proceeding of 2011 IEEE International Conference on Control System, Computing and Engineering (ICCSCE) 25–27 November 2011, (Holiday Inn, Penang, 2011), p. 571–574
5. S.A.A. Karim, B.S.M. Singh, R. Razali, N. Yahya, B.A. Karim. *Compression Solar Radiation data using Haar and Daubechies Wavelets*. In Proceeding of Regional Symposium on Engineering and Technology 2011, (Kuching, Sarawak, Malaysia, 21–23 November 2011), p. 168–174
6. S.A.A. Karim, B.S.M. Singh, B.A. Karim, M.K. Hasan, J. Sulaiman, B.J. Josefina, M.T. Ismail, Denoising solar radiation data using Meyer Wavelets. AIP Conf. Proc. **1482**, 685–690 (2012). https://doi.org/10.1063/1.4757559
7. S.A.A. Karim, *Data interpolation smoothing and approximation using cubic spline and polynomial. Book manuscript* (Springer, Berlin, 2017)
8. S.A.A. Karim, V.P. Kong. Gau*ssian Scale-Space and Discrete Wavelet Transform for Data Smoothing*. International Conference on Electrical, Control and Computer Engineering (Pahang, Malaysia, 21–22 June 2011), p. 344–348
9. M.Y. Sulaiman, W.M. Hlaing, A.M. Wahab, M.Z. Sulaiman, Analysis of residuals in daily solar radaiation time series. Renewable Energy **29**, 1147–1160 (1997)
10. Z. Sen, *Solar energy fundamentals and modeling techniques. Atmosphere, environment climate change and renewable energy* (Springer-Verlag, London, 2008)
11. T. Khatib, A. Mohamed, A, K. Sopian. A review of solar energy modeling techniques. Renewable Sustainable Energy Rev. **16**, 2864–2869 (2012)
12. Y. Wang. *smoothing splines: methods and applications (in Monographs on Statistics & Applied Probability)*, (Chapman and Hall/CRC, 2012)

13. P.C. Hansen, V. Pereyra, G. Scherer. *Least squares data fitting with applications*. (The Johns Hopkins University Press, 5 December 2012)
14. S.A.A. Karim, B.S.M. Singh, R. Razali, N. Yahya. *Data Compression Technique for Modeling of Global Solar Radiation*. In Proceeding of 2011 IEEE International Conference on Control System, Computing and Engineering (ICCSCE) 25–27 November 2011, (Holiday Inn, Penang, 2011), p. 448–35
15. J. Wu, C.K. Chan, Prediction of hourly solar radiation using a novel hybrid model of ARMA and TDNN. Sol. Energy **85**, 808–817 (2011)

# Chapter 3
# Load Profiling and Optimizing Energy Management Systems Towards Green Building Index

**Aravind CV, Ramani Kannan, I. Daniel, Alex Suresh and Sujatha Krishnan**

Due to the acute scarcity of fossil fuel resources, utilization of solar and wind power in several organizations for minimizing utility billing costs has become more significant issue. The growing attention to Distributed Generation (DG) and the use of green energy alternatives are gaining popularity as compared to conventional fossil fuel-based electricity. The optimized modelled system for the case study is to implement the functionality of an effective energy management system for the organization (UCNW building, Kuala Lumpur) using a Dual Tariff (PV/TNB Grid) system for effective energy utilization.

Energy Management System (EMS) plays a very important role on load profiling and optimization of organizational energy management systems. An optimized energy management system in this research work is developed using commercial available software and recommended results are presented.

---

Aravind CV (✉)
School of Engineering, Taylor's University, Subang Jaya, Malaysia
e-mail: aravind_147@yahoo.com

R. Kannan
Department of Electrical and Electronics Engineering, Universiti Teknologi PETRONAS, Perak Darul Ridzuan, Malaysia

I. Daniel
Department of Electrical and Electronics Engineering, Taylor's University, Subang Jaya, Malaysia

A. Suresh
Department of Electrical and Electronics Engineering, S.A Engineering College, Chennai, India

S. Krishnan
Research Associate with the Centre for Future Learning, Taylor's University, Subang Jaya, Malaysia

© The Author(s), under exclusive license to Springer Nature Singapore Pte Ltd. 2018
S. A. Sulaiman et al. (eds.), *Sustainable Electrical Power Resources through Energy Optimization and Future Engineering*, SpringerBriefs in Energy,
https://doi.org/10.1007/978-981-13-0435-4_3

## 3.1  Introduction

Electrical energy is essential for almost every household and buildings that require heating, cooling and lighting. In commercial buildings, the lighting system, HVAC system and lift operations consumes a significant amount of electrical energy that increases the importance of energy efficient operations. Our modernized lifestyle today depends solely on electrical power system from the usage of fossil fuel combined with modern power generating units to reduce cost. This is achieved by using resources, in the form of size and operation of hybrid system components, which matches its demand (energy available for utilization (kWh) and these demands need to be supplied (kW).

Energy auditing and management systems using an available commercial software tool allows users to choose different system configurations and building schematics to optimize the area of concentration and scope of buildings. Real time data analytic capabilities with advanced and accurate flow meters for measurements according to the actual amount of gas or oil that they bought from suppliers is available instantaneously in recent times [1–4]. The chapter focus at a possible designing and sizing of a complete PV system (either grid connected or standalone) based on the study conducted in a typical commercial domain, the UCSI University North Wing (UCNW). UCNW Kuala Lumpur, is geographically located at the federal territory of Malaysia with a Latitude ($3° 0862'$N) $3° 59'$N and Longitude ($101° 7411'$E) $101° 59'$E. It has an average daily solar radiation ranging from 4.994 to 5.928 kWh/m$^2$/day [5, 6] as shown in Fig. 3.1.

First, the unit for solar radiation data needs to be converted from MJm$^{-2}$ to kWhm$^{-2}$, then the Peak Sun Hour (PSH) can be calculated, when the solar

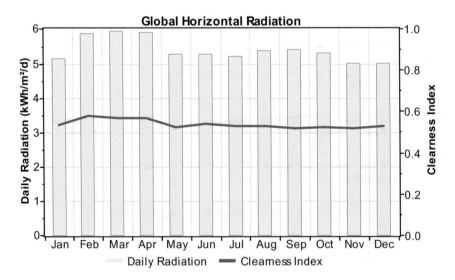

**Fig. 3.1** Annual solar radiation tabulation

irradiance is $1\,\text{kWhm}^{-2}$. For example, on Jan 2008 in Pulau Pinang, the solar radiation data is $13.4671\,\text{MJm}^{-2}$. The conversion factor is given by:

$$1\frac{\text{MJ}}{\text{m}^2} \times \frac{1\,\text{kWh}}{3.6\,\text{MJ}} = 3.7408\,\text{kWhm}^2 \tag{3.1}$$

Therefore, to produce a 3.7408 kWh of energy per $\text{m}^2$, the radiation would have to continue at the rate of $1\,\text{kWm}^{-2}$ for 3.74 h. Thus, the number of PSH is 3.74.

## 3.2 Energy Management Systems

Energy management system includes auditing through three ways including walk-through [6], standard methods [7] and through computer simulation auditing tools. The walk through involves a tour of the facility (like UCNW, Malaysia),

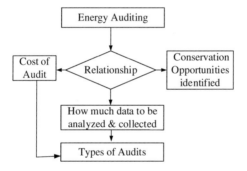

**Fig. 3.2** Stages of energy auditing

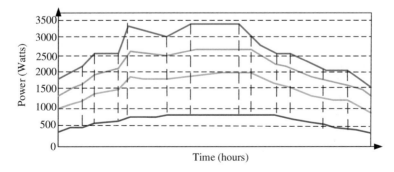

**Fig. 3.3** UCNW 24-hourly power (kW) load data profile on a selected P.T. Ratio system of 11 kV/110 V on March 30, April 2, 5, 17, 2011 (30-minute Interval)

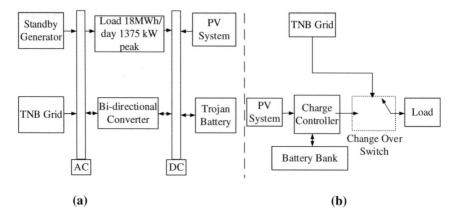

**(a)**                                                    **(b)**

**Fig. 3.4** **a** overview of the possible Hybrid Renewable Energy System (HRES), **b** Integration of possible to the TNB configuration (PV/TNB) for UCNW

economic analysis of recommending conservative measure. Figure 3.2 shows the energy auditing flowchart approach used in this research.

Figure 3.3 shows the hourly load data that used to design the existing system. Hybrid Renewable Energy System (HRES) consist of generators, batteries, and solar panel (PV). HOMER is used for designing of hybrid systems and most hybrid systems are made of Solar, Wind and Hydro Resource and it is used to derive estimated cost for dual-tariff PV system, typically as shown in Fig. 3.4. This acts as a platform for performing energy analysis aimed at the identification of possible energy saving measure [7]. HOMER, is used for the optimization of the energy system for UCNW. PSCAD, commercial industrial standard tool is used to optimize the sizing and evaluation of the system proposed.

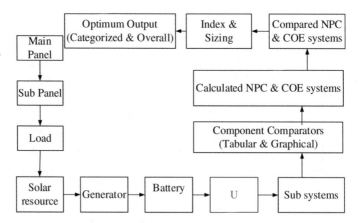

**Fig. 3.5** Block diagram of HRES interface console

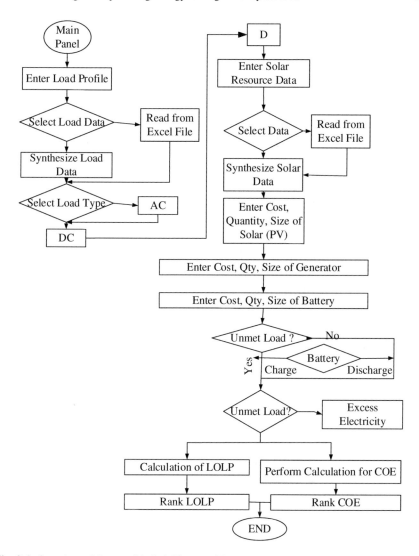

**Fig. 3.6** Overview of the possible hybrid renewable energy system

## 3.3 Methods

The method used in this investigations consists of four steps; simulation, optimization, sensitivity analysis and analysis on the findings. First, the simulation evaluates the performance by simulating many different system configurations to determine the technical feasibility and excessive electricity and life cycle cost of the system. Secondly, optimization performed for different system configurations to achieve the lowest life-cycle cost. Finally, the sensitivity analysis process performs

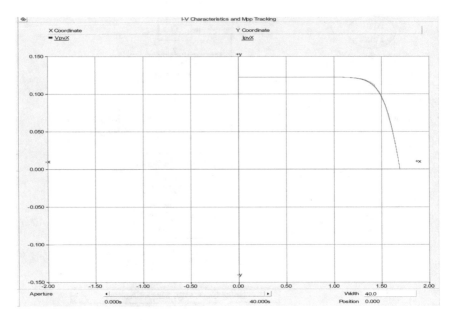

**Fig. 3.7** The UCNW I-V characteristics and MPPT graph in the proposed system (based on incremental conductance algorithm)

optimization under a range of input assumptions to assess the effect of uncertainty that allow us to evaluate the changes in the variables such as the average wind speed. With the data, the process starts where the data segregated for other sub-systems. The optimized data through sensitivity analysis is as shown in Fig. 3.5. Figure 3.6 shows the block representation for the proposed system under investigations. The PSCAD proposed system evaluate the filter design and harmonic analysis, optimal design of controller parameters, and developing system coding as shown in Fig. 3.7.

## 3.4   Results and Discussions

Graphical analysis of a 12-month period of the energy usage system of the building presented. An efficient energy management system must comprise of energy auditing, monitoring and management with capabilities of scheduling, duty cycling,

**Table 3.1** Summary of the optimal HRES

| | Optimum system (Case 1) 5882 kWh/day | Optimum system (Case 1) 6113 kWh/day | | Optimum system (Case 1) 6160 kWh/day |
|---|---|---|---|---|
| *Case 1* | | | | |
| Sizing of conventional generators (kW) | 20 | 20 | | 20 |
| Sizing of renewable (PV Panel) (kW) | 9 | 100.8 | | 190 |
| Size of converter | 5 | 31 | | 31 |
| Number of batteries | 0 | 0 | | 0 |
| *Case 2* | | | | |
| Sizing of conventional generators (kW) | 20 | 20 | | 20 |
| Sizing of renewable (PV Panel) (kW) | 9 | 100.8 | | 190 |
| Size of converter | 5 | 31 | | 31 |
| Number of batteries | 32 | 32 | | 32 |
| | Optimum system (Case 3) 5882 kWh/day | Optimum system (Case 3) 6113 kWh/day | Optimum system (Case 3) 6160 kWh/day | System with derating factor of 0.7895 (5800 kWh/day) |
| Sizing of conventional generators (kW) | 20 | 20 | 20 | 0 |
| Sizing of renewable (PV Panel) (kW) | 21 | 190 | 190 | 9 |
| Size of converter | 5 | 31 | 31 | 1 |
| Number of batteries | 0 | 0 | 64 | 0 |

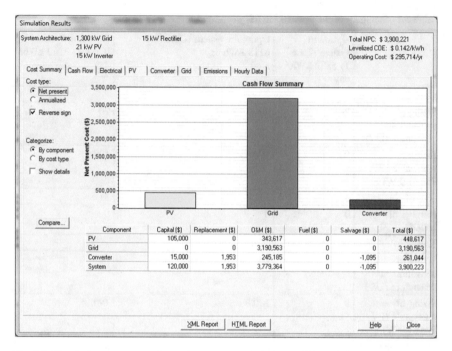

**Fig. 3.8** Detailed cost summary for each component in the Solar-PV tariff system for the UCNW Malaysia for 5882 kWh/day

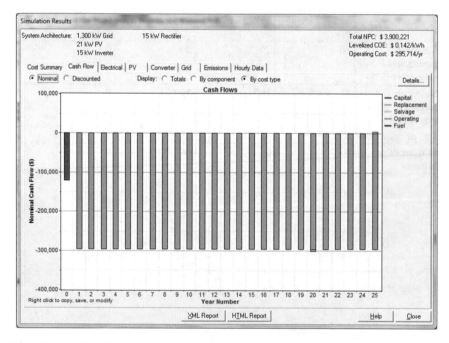

**Fig. 3.9** Cash flows in 25 years

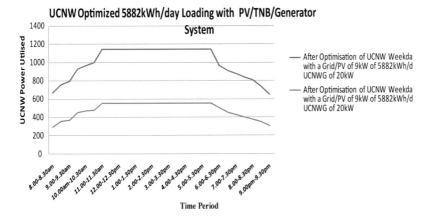

**Fig. 3.10** Optimized output for the possible UCNW (5882 kWh/day)

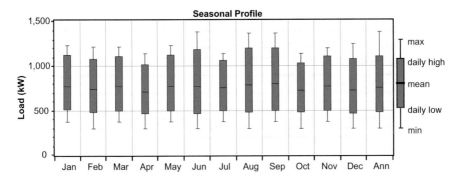

**Fig. 3.11** Annual average load for the UCNW Malaysia

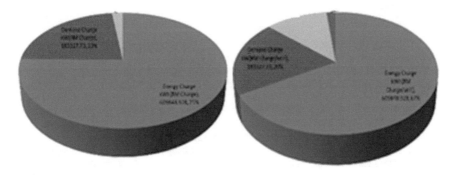

**Fig. 3.12** Effect of UCNW electricity expense without and with a PV/Grid system

demand limiting, optimal starting, monitoring and direct digital control. In terms of the economical aspect of analysis, an estimated cost for PV/Grid system using HOMER is developed. Table 3.1 shows the summary of the optimal HRES. The possible way of optimal configurations suggested based on the study is explicitly seen. The cost includes the capital, replacement, operation and maintenance (O & M) as in Fig. 3.8. The cash flow analysis as seen in Figs. 3.9 and 3.10 suggest the possible optimum system of UCNW (5882 kWh/day). Figure 3.11 shows the annual average load for the UCNW Malaysia and Fig. 3.12 shows the effect of UCNW electricity expense without and with a PV/Grid system.

## 3.5 Summary

This chapter can be summarized as follows:

1. The proposed setting to shut of the chilled water supply into the AHU during the lunch mode observed as feasible as the number of staff who would remain in the office would be less than 10%. The estimated saving in the annual cooling energy was 33.368 kWh or about 1.7%. Since no capital investment is required, the approach is practical.
2. Increasing the set point temperature during the rain considered as an opportunity for a place that is blessed with consistent rain throughout the year. In the present study, by raising the set point temperature by 2 °C when it rains would reduce the annual cooling energy by approximately 370,000 kWh or 36%.
3. Optimization of lightings in a lecture room practiced since it may contribute in saving energy. In the present work a 50%, reduction in the cooling load observed if the number of occupants is small (20 persons).

## References

1. EIA. Energy Information Administration—EIA— Official Energy Statistics from the U.S. Government. Available: www.eia.doe.gov. (2010, 29 January)
2. T.K. Soon, S. Mekhilef, A fast-converging mppt technique for photovoltaic system under fast-varying solar irradiation and load resistance. IEEE Trans. Industr. Inf. 11(1), 176–186 (2015)
3. A. Belhamadia, M. Mansor, M.A. Younis. *Assessment of wind and solar energy potentials in Malaysia*. Proceedings of the IEEE Conference on Clean Energy and Technology (CEAT), (Lankgkawi, 2013), p. 152–157. https://doi.org/10.1109/ceat.2013.6775617
4. M.H. Alsharif, R. Nordin, M. Ismail, *Optimization of hybrid renewable energy power system for urban LTE base station deployment in Malaysia*. Proceedings of the Second IEEE International Symposium on Telecommunication Technologies (ISTT), (Langkawi, 2014), p. 1–5. https://doi.org/10.1109/istt.2014.7238166
5. A. Arya, Jyoti, P. Arunachalam. *Review on industrial audit and energy saving recommendation in aluminum industry*. Proceedings of the International Conference on Control,

Instrumentation, Communication and Computational Technologies (ICCICCT), (Kumaracoil, 2016), p. 758–764. https://doi.org/10.1109/iccicct.2016.7988054

6. J. Chabarek, P. Barford. *Energy Audit: Monitoring power consumption in diverse network environments.* Proceedings of the International Green Computing Conference Proceedings, (Arlington, VA, 2013), p. 1–10. https://doi.org/10.1109/igcc.2013.6604505

7. R. Saidur, M. Hasanuzzaman, M.M. Hassan, H.H. Masjuki, overall thermal transfer value of residential buildings in Malaysia. Journal of Applied Sciences **9**, 2130–2136 (2009)

# Chapter 4
# Planning of Distributed Renewable Energy Resources Using Genetic Algorithm

**Nursyarizal Mohd Nor, Abid Ali, Taib Ibrahim and Mohd Fakhizan Romlie**

Due to the limitations of fossil fuel reserves and environmental issues, distributed renewable energy resources (DRER) have become a good alternative solution for producing electricity. In addition to supply of clean energy, the integration of DRER in distribution networks helps to minimize power losses and also to enhance bus voltage profiles. The number, locations, and sizes of the DRER significantly affect the quality of the electrical networks. Therefore, determination of appropriate sizes and locations of DRER has become one of the main goals for power operators. Aiming at this problem, this chapter presents a multi-objective optimization for the planning of Distributed Renewable Energy Resources (DRER) in the distribution networks. The objective functions are to minimize power losses, enhance bus voltage profiles and minimize the total size of DRER units. A weighted sum approach by Mixed Integer Optimization with Genetic Algorithm (MIOGA) is used to solve the multi-objective optimization problem. To verify the effectiveness of the proposed method, simulation studies are conducted on IEEE 33 buses and 69 buses test distribution networks. The results obtained by the proposed method show that the installation of multiple DRER at appropriate locations provides more benefits in terms of reductions in power losses and enhancement of bus voltage profiles. The proposed method is seen effective in solving the multi-objective optimization problem for the planning of distributed renewable energy resources in distribution networks.

N. M. Nor (✉) · A. Ali · T. Ibrahim · M. F. Romlie
Department of Electrical and Electronics Engineering, Universiti Teknologi PETRONAS, Bandar Seri Iskandar, 32610 Seri Iskandar, Perak, Malaysia
e-mail: nursyarizal_mnor@utp.edu.my

© The Author(s), under exclusive license to Springer Nature Singapore Pte Ltd. 2018
S. A. Sulaiman et al. (eds.), *Sustainable Electrical Power Resources through Energy Optimization and Future Engineering*, SpringerBriefs in Energy,
https://doi.org/10.1007/978-981-13-0435-4_4

## 4.1 Introduction

Unlike fossil fuel-fired thermal power plants, small-scale power stations that utilize renewable energy sources are gaining more attention among the power producers. Distributed generations as well as distributed renewable energy resources (DRER) are small electrical generators, and are preferred to be installed to loads [1]. The DRER, which use sustainable energy sources to produce electricity, like solar photovoltaic modules or wind turbines, are considered more advantageous. This is mainly because the renewable energy sources are environment friendly and are expected to produce low-cost electricity in coming years [2]. Among the various renewable energy sources, solar and wind energy are getting more attention in recent times and are considered to be the dominating sources for generation of electricity in the near future [3].

Based on the electrical characteristics in terms of real and reactive power delivering capability, the DRER can be classified into four major types [4]:

(1) Type 1: DRER are capable of injecting active power (P) only.
(2) Type 2: DRER are capable of injecting reactive power (Q) only.
(3) Type 3: DRER are capable of injecting both active power (P) and reactive power (Q).
(4) Type 4: DRER are capable of injecting active power (P) and are capable of consuming reactive power (Q).

Photovoltaic (PV), micro turbines, and fuel cells are some examples of Type 1 DRER, whereas, synchronous compensators such as gas turbines, capacitor banks, etc. are considered as Type 2 DRER. Synchronous machines such as co-generation, gas turbine, etc. are categorized as Type 3 DRER, whereas, wind turbines, which use induction generators, are regarded as Type 4 DRER.

Since the acceptance of renewable energy for production of electricity at grid level, research related to the sizing and placement of the DRER units has become one of the most important topics in the field of power system planning. Since sizing and locations of the DRER units have significant influence on the quality of grid [5], installation of DRER units in the distribution networks is done based on the consideration of several network constraints i.e. reduction in system losses and improvements in bus voltage profiles. In the literature, various optimization techniques and system parameters for determining the optimum sizes and locations of DRER units in distribution networks had been proposed. The benefits and limitations of different optimization techniques used for the sizing and placement of DRER units in distribution networks are documented in the reports by Georgilakis and Hatziargyriou [6] and Tan et al. [7].

Most studies related to the utilization of DRER in distribution network only considered the supply of active power. Moreover, results of various studies revealed that as compared to installation of single DRER unit only, the installation of multiple DRER units could bring more benefits to distribution networks. For minimization of network power losses and to improve voltage stability, the sizing and locating of multiple DRER units were performed by using Bacterial Foraging Optimization Algorithm (BFOA) [8], a novel combination of Genetic Algorithm (GA) and Particle Swarm Optimization (PSO), also known as the GA/PSO method [9], and Loss Sensitivity Factor Simulated Annealing (LSFSA) [10]. The research work by Prabha and Jayabarathi [11] employed Invasive Weed Optimization (IWO) approach to optimize the sizing of multiple DRER unit; they used Loss Sensitivity Factor (LSF) method to determine the locations of DRER units. The sizing and placement of multiple DRER units were also performed by using Dynamic Adaptation of Particle Swarm Optimization (DAPSO) [12], Teaching Learning Based Optimization (TLBO) technique [13], Harmony Search Algorithm (HSA) [14], and analytical approach [15]. The methodology for DRER sizing and placement in most studies used IEEE 33 bus and IEEE 69 bus distribution networks. It is known that installation of single DRER can reduce more than 60% of power loss in the distribution networks.

The multi-objective optimization method could provide solutions for several parameters instead of a single one. The sizing and placement of DRER units by considering multiple objective functions has been reported in several studies. For instance, a multi-objective Shuffled Bat algorithm reported by Yammani et al. [16] was proposed to evaluate the impact of DRER placement and sizing in distribution system by considering the power losses, cost and voltage deviation as objective functions. A multi-objective function optimization in the work of Karami et al. [17] through the use of genetic algorithm was proposed by considering the system load-ability, and total costs that include investment costs of distributed generations and distributed static compensators and network loss. Furthermore, multi-objective optimization of DRER placement and sizing using similar objective functions by using Hybrid Big Bang-Big Crunch [18], Pareto Frontier Differential Evolution (PFDE) algorithm [19], advanced Pareto-front non-dominated sorting multi-objective particle swarm optimization (Advanced-PFNDMOPSO) [20], sequential quadratic programming deterministic technique [21] and Chaotic Artificial Bee Colony (CABC) algorithm [22] were also reported. The sizing and placement of DRER units with multiple objectives were performed by using different heuristic optimization techniques. For installations of DRER, different load models are used. Objective functions considered in different studies were minimization of power losses, enhancement of bus voltage profiles, voltage stability index, and minimization of cost of the system etc.

From the literature review, it is revealed that most studies used Pareto front with the support from a set of fuzzy expert rules for optimization of multiple objectives. The problem with this approach is, firstly, the set of Pareto front optimization solutions is determined and then the final decision is made according to the preferences of the objective functions. The computational efforts to solve multi-objective problem using Pareto front increase significantly with the number of the objective functions [23]. On the other hand, weighted sum (WS) is considered as an efficient method to solve the optimization of multiple objectives with respect to preference of each objective function [24]. It is a classical approach to solve a multi-objective problem, which converts the multi-objective problem into a single objective problem. The main advantage of this approach is a straightforward implementation. In addition, this approach is computationally efficient [25]. This approach was adopted in various studies including sizing and placement of DRER in distributed networks [26–28]. The details and formulations of converting multi-objective problems into single objective problem are available in the report by Deb [29].

Heuristic methods are usually considered robust and they provide near-optimal solutions for large and complex problems [30]. Among the various heuristic methods, Genetic Algorithm (GA) is well known for solving the complex optimization problems due to its powerful search and optimization capabilities [31, 32]. The GA based method is used by Kim et al. [33] to determine the size and location of DRER unit. Moradi et al. [34] proposed a hybrid method based on Imperialist Competitive Algorithm (ICA) and Genetic Algorithm (GA) for optimal placement and sizing of DRER units and capacitor banks, simultaneously. Gandomkar et al. [35] examined the multi-objective based placement and penetration level of DRER units by using Genetic Algorithm (GA) combined with Multi-Attribute Decision Making (MADM) method.

Mixed Integer Optimization with Genetic Algorithm (MIOGA) is considered more efficient in finding out best solutions than the simple GA. Numerical comparison experiments in the work by Yokota et al. [36] clearly demonstrated the efficiency of the MIOGA method. Therefore, the proposed approach in this chapter adopts a weighted sum approach by Mixed Integer Optimization with Genetic Algorithm (MIOGA) to solve the multi-objective optimization for the planning of DRER in distribution networks.

## 4.2 Study Objective and Methodology

This work is intended to determine the optimum sizes and placements of Distribution Renewable Energy Resources (DRER) using a weighted sum approach by Mixed Integer Optimization with Genetic Algorithm (MIOGA). The objective functions considered for the study are minimization of the power losses and enhancement of the bus voltage profiles. The review of literature reveals that as compared to installation of single DRER unit, the installation of multiple DRER units in the distribution networks could provide more benefits in terms of reduction in power losses. Therefore, the proposed method in this study is used to determine the optimum sizes and locations of single, two and three DRER units in the distribution networks.

To investigate the effectiveness of the proposed method, this research study uses four voltage dependent load models, which include constant, industrial, residential and commercial load models. The proposed algorithm is tested on IEEE 33 bus and IEEE 69 bus test distribution networks and multi-objective optimization problem is solved. The simulation code of the proposed method is developed in MATLAB 2015a by using Intel® Core™ i5 CPU 3.25 GHz with 4 GB of RAM.

## 4.3 Modeling of the System Parameters

### 4.3.1 Power Flow Equations

Power flow analysis is a basic and necessary tool for electrical systems to determine the exact electrical performance under the steady state conditions. The load flow analysis provides the real (kW) and reactive power (kVAr) losses across the branches and voltage magnitudes and angles at different nodes of the electrical network. This analysis is necessary to perform during power system planning, operation, optimization and control for integration of distribution generations in distributed networks.

Some of the basic power flow analysis methods used in electrical transmission networks are Newton Raphson (NR), Gauss Seidel (GS) and Fast Decoupled Load Flow [37]. However, these methods become inefficient for the distribution network due to their radial structure, high R/X ratio and unbalanced loads [38]. Alternative to this, load flow analysis in distribution network is performed by other computational methods. In the literature, several approaches for power flow analyses in distribution networks have been developed. The power flow analysis by backward/forward sweep method using basic formulations of Kirchhoff's laws was proposed by Shirmohammadi et al. [39]. The analytical method for solving radial distribution networks using simple algebraic expression [40] and approximate solution of the nonlinear power equations for a balanced power distribution network [41] was formulated. Haque et al. [42] presented a power flow analysis method for meshed networks. A solution for power flow analysis in distribution system using bus injection to branch current matrix and branch current to bus voltage matrix was presented by Teng [43]. Among these approaches, the backward/forward sweep method is commonly used due to its computational efficiency and solution accuracy [44–48]. In this chapter, standard backward/forward sweep method is used for load flow analysis in the distribution networks.

The backward/forward sweep method includes two steps: the backward sweep and the forward sweep. In backward sweep, voltage and currents are computed using Kirchhoff's Voltage Law (KVL) and Kirchhoff's Current Law (KCL) from the farthest node to the source node. In forward sweep, the voltage at each bus is then updated starting from source node to the farthest node.

The single line diagram of a section of the distribution network is shown in Fig. 4.1, which comprises two buses; k and (k + 1), connected through a branch line i. The flows of power in this section can be computed by [15]:

$$P_{(i)} = P_{D(k+1)} + P_{loss(i)} \tag{4.1}$$

$$Q_{(i)} = Q_{D(k+1)} + Q_{loss(i)} \tag{4.2}$$

$$V_{(k+1)} = V_{(k)} - I_{(i)}\left(R_{(i)} + jX_{(i)}\right) \tag{4.3}$$

**Fig. 4.1** One-line diagram of a two-bus section in a radial distributed network

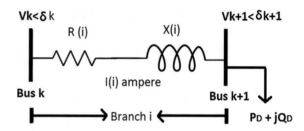

Resistance and reactance of the branch i are represented by Ri and Xi, respectively. $I_{(i)}$ is the current that is flowing through branch i. $P_{(i)}$ and $Q_{(i)}$ are the active and reactive powers that flow across branch i. $P_{D(k+1)}$ and $Q_{D(k+1)}$ are active and reactive loads, connected at bus (k + 1).

The total power injection at bus (k + 1) is the sum of total loads connected at bus (k + 1) and the effective power which is flowing towards bus (k + 1) from the other buses. The two powers [11] can be written as:

$$P_{(i)} = P_{(k+1),\text{eff}} + P_{D(k+1)} + P_{\text{loss}(i)} \tag{4.4}$$

$$Q_{(i)} = Q_{(k+1),\text{eff}} + Q_{D(k+1)} + Q_{\text{loss}(i)} \tag{4.5}$$

The power losses [49] across branch i can be calculated by:

$$P_{\text{loss}(i)} = R_{(i)} * \frac{P_{(k+1)^2} + Q_{(k+1)^2}}{\left|V_{(k+1)}\right|^2} \tag{4.6}$$

$$Q_{\text{loss}(i)} = X_{(i)} * \frac{P_{(k+1)^2} + Q_{(k+1)^2}}{\left|V_{(k+1)}\right|^2} \tag{4.7}$$

where, $P_{\text{loss}(i)}$ and $Q_{\text{loss}(i)}$ are the active and reactive power losses across branch i.

## 4.3.2   Power from DRER in to Grid

The type of DRER which can supply active power (Type 1) is considered for this study. The output power from the DRER units can be used as negative load [50]. Assuming the DRER unit is installed at bus (k + 1), the total active and reactive power injections for bus (k + 1) can be calculated by including active power ($P_{\text{DRER}}$) and reactive power ($Q_{\text{DRER}}$) from DRER into Eqs. (4.4) and (4.5). This can be expressed as:

$$P_{(i)} = P_{(k+1),\text{eff}} + P_{D(k+1)} + P_{\text{loss}(i)} - P_{\text{DRER}} \tag{4.8}$$

$$Q_{(i)} = Q_{(k+1),\text{eff}} + Q_{D(k+1)} + Q_{\text{loss}(i)} - Q_{\text{DRER}} \tag{4.9}$$

Similarly, the power losses across the branch i after the installation of DRER at bus (k + 1) can be calculated by including active power ($P_{\text{DRER}}$) and reactive power ($Q_{\text{DRER}}$) from DRER into Eqs. (4.6) and (4.7). This can be expressed by:

$$P_{\text{loss}(i)}^{\text{DRER}} = R_{(i)} * \frac{(P_{(k+1)} - P_{\text{DRER}})^2 + (Q_{(k+1)} - Q_{\text{DRER}})^2}{|V_{k+1}|^2} \tag{4.10}$$

$$Q_{\text{loss}(i)}^{\text{DRER}} = X_{(i)} * \frac{(P_{(k+1)} - P_{\text{DRER}})^2 + (Q_{(k+1)} - Q_{\text{DRER}})^2}{|V_{(k+1)}|^2} \tag{4.11}$$

### 4.3.3   Load Models

In practical distribution networks, the load can be classified as constant, industrial, residential and commercial loads [49, 51, 52]. These types of loads are also referred to as voltage dependent load models. Depending on the type of loads, the new loads $P_{\text{Dnew}}$ for each bus are calculated by:

$$P_{\text{Dnew}(k)} = P_{\text{Da}(k)} * V_{(k)}^{v\alpha} \tag{4.12}$$

$$Q_{\text{Dnew}(k)} = Q_{\text{Da}(k)} * V_{(k)}^{vr} \tag{4.13}$$

where $P_{\text{Da}}$ and $Q_{\text{Da}}$ are the actual active and reactive loads connected at bus k, and V(k) is the voltage magnitude of $k^{\text{th}}$ bus which is calculated under base load conditions. The two parameters $v\alpha$ and $vr$ are known as voltage coefficients for active and reactive power loads. The values of $v\alpha$ and $vr$ for constant, industrial, residential and commercial loads are provided in Table 4.1 [49, 51, 52].

| Table 4.1  Voltage coefficients for active and reactive power loads | Type of load | $v\alpha$ | $vr$ |
|---|---|---|---|
| | Constant | 0 | 0 |
| | Industrial | 0.18 | 6.00 |
| | Residential | 0.92 | 4.04 |
| | Commercial | 1.51 | 3.40 |

### 4.3.4 Problem Formulation

The main objective of this study is to reduce the total power losses in the distribution networks. The total power losses in distribution network can be calculated by summing the active and reactive power losses across all the branches in the network [14].

$$P_{loss\_total} = \sum_{i=1}^{no.of\ branches} P_{loss(i)} + jQ_{loss(i)} \qquad (4.14)$$

Similarly, with the addition of the DRER units, the total power losses can be calculated by:

$$P_{loss\_total}^{DRER} = \sum_{i=1}^{no.of\ branches} P_{loss(i)}^{DRER} + jQ_{loss(i)}^{DRER} \qquad (4.15)$$

The deviations in the network losses due to the installation of DRER units can be observed by dividing the real parts of Eq. (4.14) with the real parts of Eq. (4.15). This method is also termed as Total Power Loss Index (TPLI) [52] and mathematically it can be expressed as:

$$\text{Total Power Loss Index (TPLI)} = \sum \frac{\text{real}\left(P_{loss\_total}^{DRER}\right)}{\text{real}\left(P_{loss\_total}\right)} \qquad (4.16)$$

Minimization of TPLI is the first objective function for this study and this can be expressed as:

$$f1 = \text{minimization [Total Power Loss Index (TPLI)]} \qquad (4.17)$$

In addition, to reduce the network power losses, the addition of DRER units in distribution network improves the bus voltage profiles. The change in bus voltage magnitude is measured as deviation in bus voltages [15], which can be calculated by:

$$VDev_k = |1 - V_k|^2 \qquad (4.18)$$

In voltage dependent load models, the bus voltages starts to decrease with the increase in loads values and electrical networks experience the minimum voltages during the peak load hours [53]. Using Eq. (4.18), the bus with lowest voltage magnitude will give maximum voltage deviation ($VDev_{max}$). In that case, the bus with maximum voltage deviation ($VDev_{max}$) will be given more importance as the voltage deviation magnitude of this bus will indicate the status of whole network.

Therefore, in addition to reduce the network losses, this study puts a continuous monitoring on the magnitudes of $VDev_{max}$. The maximum voltage deviation ($VDev_{max}$) [8] can be calculated by using following relationship:

$$VDev_{max} = max(|1 - V_k|^2) \qquad (4.19)$$

The second objective of this study is to minimize the maximum voltage deviation values determined by Eq. (4.19). Furthermore, this can be written as the following expression.

$$f2 = minimization\ [VDev_{max}] \qquad (4.20)$$

From the review of literature, it is known that 100 kW of power losses in IEEE 33 bus distribution network can be reduced by installing a DRER of 2.59 MW in capacity [5]. However, to investigate the impact of others sizes of DRER on the reduction of power losses, a simulation was run using different sizes of DRER. Through this investigation, it was revealed that about 91 kW of power losses in the same network can be reduced by installing a DRER of 1.48 MW capacity. This shows that the rate of reduction in power losses due to addition of DRER in distribution network is not proportional to the size of DRER unit. To reduce the size of DRER, an additional constraint is used. Thus, the third objectives function of this study is the minimization of total size of the DRER units:

$$f3 = minimization\left[ \sum_{i=1}^{Total\ no.\ of\ DRERs} DRERi \right] \qquad (4.21)$$

### 4.3.5  Network Constraints

To solve the multiple objective functions problem, the simulation of the proposed algorithm must fulfil the following constraints.

(a)  Distributed Generation Capacity

$$P_{DG}^{min} < P_{DG} < P_{DG}^{max} \qquad (4.22)$$

Here $P_{min}^{DRER}$ is equal to 0 and $P_{max}^{DRER}$ is equal to the peak active power load of each voltage dependent load. Therefore, $P_{max}^{DRER}$ for each type of load type will not be the same.

(b) Network Power Balance

$$P_{substation} + P_{DRER} = P_{DTotal} + P_{loss\_total} \qquad (4.23)$$

$$Q_{substation} + Q_{DRER} = Q_{DTotal} + Q_{loss\_Total} \qquad (4.24)$$

$P_{substation}$ and $Q_{substation}$, are, respectively, the active and reactive power supplies to distribution network from the substation.

(c) Locations for DRER

Since Bus 1 in the distributed network is considered as a slack bus, the DRER units are proposed to be connected to any bus from Bus 2 to the last bus in the network:

$$2 \leq DRER_{location} \leq \max (\text{number of buses}) \qquad (4.25)$$

(d) Bus Voltage Limits

$$V_{min} \leq V \leq V_{max} \qquad (4.26)$$

In order to maintain the power quality of distribution network, the bus voltage magnitude will remain under a $V_{max}$ of 1.0 p.u.

### 4.3.6 Multi-objective Indices

The distribution networks are characterised for their high R/X ratio that could cause considerable power losses along the feeders. It is well-known that the percentage of power losses at distribution networks is higher than the percentage of power losses in transmission networks. For example, the power losses in Tenaga Nasional Berhad's (TNB) distribution networks in 2015 were 6.21%, as compared to power losses in its transmission networks of 1.47% [54]. Consequently, reduction in power losses at distribution level has become one of the greatest challenge to power distribution utilities.

To solve the multi-objective optimization, weighting values (w) were assigned to each objective function. These weighting values are intended to give the relative importance for each objective function. The proper weighting values depend on the experience and priorities set by system planner. The installation of DRER has a significant impact on the power losses and bus voltage profile as well. However, due to the impact on revenues of the distribution utilities, the minimization of power losses is currently given more importance than enhancement of bus voltage profile [19, 23, 52, 55]. In the current study, the chosen weight for minimization of power losses is 0.6, giving it maximum priority and leaving the remaining two objective functions to 0.2 each. To include all the indices in the analysis, a multi-objective index (IMO) can be defined as a combination of the all objective function f1, f2 and f3 with chosen weights for w1, w2 and w3.

$$IMO = (w1 * f1) + (w2 * f2) + (w3 * f3) \qquad (4.27)$$

where,

$$\sum_{i=1}^{\text{no. of obj. functions}} wi = 1 \qquad (4.28)$$

## 4.4 Simulation Results and Discussions

This section consists of results of simulation, calculated by the Mixed Integer Optimization with Genetic Algorithm (MIOGA) for solving the multi-objective problem using weighted approach for the sizing and placement of DRER units in the distribution networks.

The Genetic Algorithm (GA) is a heuristic optimization technique used to solve constrained and unconstrained problems. Inspired by Darwinian's principle, this optimization technique follows the biological growth process to solve real world problems by using an evolution and a natural selection [56, 57]. GA holds a data structure similar to chromosomes. The chromosomes are changed by using the proper values of different parameters such as selection, crossover and mutation operators. The details of GA and its parameters are provided elsewhere [9, 58].

Before the running of GA, the load flow analysis is run to calculate the power losses and bus voltage profiles without DRER for each load model. This process is also known as base case. Then GA is run, which randomly chooses the size and locations for DRER units. The load flow analysis is run again to calculate the power losses and bus voltage profiles by including the DRER. The values of power losses and bus voltage profiles with DRER are compared with values of power losses and bus voltage profiles without DRER. The process keeps running until optimum results are achieved. The optimization process flowchart of GA is provided in Fig. 4.2.

The proposed algorithm is simulated and applied on two test systems; IEEE 33 bus and IEEE 69 bus distribution networks. The active and reactive loads of the two networks are 3.71 MW & 2.3 MVAr, and 3.8 & 2.69 MVAr, for the IEEE 33 bus and IEEE 69 bus, respectively. The total real and reactive losses of the two test systems in base case system are known to be 211 kW & 143 kVAr for the IEEE 33 bus, and 225 kW & 102 kVAr for the IEEE 69 bus [59–61]. The weakest buses of the IEEE 33 and IEEE 69 test networks are Bus 18 and Bus 65, with voltage magnitudes of 0.904 p.u. and 0.909 p.u., respectively. The single line diagrams of

**Fig. 4.2** Flowchart for
DRER sizing and placement

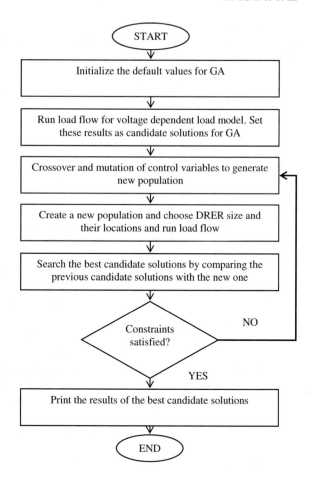

the two distribution networks are provided in Figs. 4.3 and 4.4, whereas, the line and load data of the two distribution networks are available in the report by Venkatesh et al. [62].

The first bus in distribution network is considered as slack bus, and this bus cannot be used for installation of DRER. Therefore, DRER can be installed at any bus from Bus 2 to 33 in IEEE 33 bus network and from Bus 2 to 69 in IEEE 69 bus network. For both test cases, four types of voltage dependent load models are used, which include constant, industrial, residential and commercial loads. The four voltage dependent loads are modelled by using Eqs. (4.12) and (4.13). The constraints for simulations for the both test cases with single, multiple DRER units are kept the same as provided earlier.

The simulation results of the four voltage dependent load models, having a single DRER unit in IEEE 33 bus and IEEE 69 bus test cases are summarized in Tables 4.2 and 4.3. For both the test cases, the size of DRER unit for each load model is considerably different from each other. This is mainly due to the difference

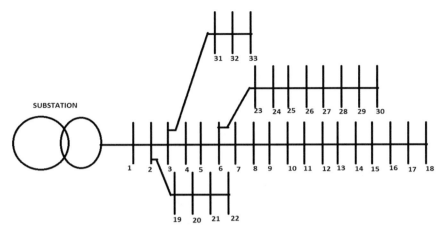

**Fig. 4.3** Single-line diagram of IEEE 33 bus distribution network

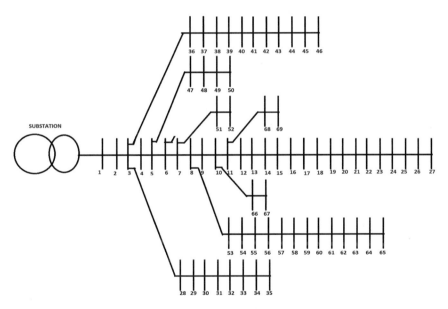

**Fig. 4.4** Single-line diagram of IEEE 69 bus distribution network

of energy consumption for each load model during the peak load time. The optimum locations of the DRER units are Bus 8 and Bus 61 in IEEE 33 bus and IEEE 69 bus distribution network, respectively.

The minimum size of DRER unit in IEEE 33 bus test case is found for commercial load model and with installation 1.59 MW size DRER unit, 46.4% of the total power losses are reduced. On the contrary, the maximum size of DRER in the

**Table 4.2** Summary of results with 1 DRER in IEEE 33 bus test system

| Parameters | Constant | | Industrial | | Residential | | Commercial | |
|---|---|---|---|---|---|---|---|---|
| | Without | With DRER | Without | With DRER | Without | With DRER | Without | With DRER |
| DRER size (MW)@Bus no | | 1.63@8 | | 1.65@8 | | 1.62@8 | | 1.59@8 |
| Total power losses (kW) | 211.00 | 118.83 | 163.69 | 77.83 | 159.12 | 81.78 | 152.63 | 81.81 |
| Reduction in power loss (%) | | 43.68% | | 52.46% | | 48.60% | | 46.40% |
| Total power loss index | | 0.56 | | 0.48 | | 0.51 | | 0.54 |
| Voltage deviations | 0.134 | 0.042 | 0.100 | 0.024 | 0.099 | 0.025 | 0.095 | 0.024 |
| Min V@Bus | 0.904@18 | 0.941@33 | 0.917@18 | 0.954@33 | 0.918@18 | 0.953@33 | 0.92@18 | 0.953@33 |

**Table 4.3** Summary of results with 1 DRER in IEEE 69 bus test system

| Parameters | Constant | | Industrial | | Residential | | Commercial | |
|---|---|---|---|---|---|---|---|---|
| | Without | With DRER | Without | With DRER | Without | With DRER | Without | With DRER |
| DRER size (MW)@Bus no | | 1.67@61 | | 1.67@61 | | 1.62@61 | | 1.59@61 |
| Total power losses (kW) | 224.95 | 84.69 | 171.42 | 42.45 | 164.91 | 50.31 | 156.97 | 53.20 |
| Reduction in power loss (%) | | 62.35% | | 75.24% | | 69.49% | | 66.11% |
| Total power loss index | | 0.38 | | 0.25 | | 0.31 | | 0.34 |
| Voltage deviations | 0.099 | 0.024 | 0.080 | 0.017 | 0.076 | 0.017 | 0.072 | 0.016 |
| Min V@bus | 0.909@65 | 0.967@27 | 0.92@65 | 0.971@27 | 0.922@65 | 0.971@27 | 0.924@65 | 0.971@27 |

same network is revealed for industrial load model and 52.46% of the total power losses are reduced with the installation of 1.65 MW size DRER unit.

The improvement in lowest voltage bus due to the installation of DRER is also noticed. Without the DRER in IEEE 33 bus, the maximum voltage deviation of 0.134 p.u. is observed in constant load model, whereas with the addition of the DRER unit, this value is reduced to 0.042 p.u. This is very significant in term of enhancement in bus voltage profile. After the installation of DRER unit, the weakest bus of the network is Bus 18, but the installation of DRER unit helped to enhance its voltage magnitude from 0.904 p.u. to 0.941 p.u.

The minimum size of DRER unit in IEEE 69 bus test case is also found for commercial load model, and with the installation of a 1.59 MW size DRER unit, 66.11% of total power losses are reduced. Whereas, the maximum size of DRER in this network is found for industrial load model and 75.24% of the total power losses are reduced with the installation of a 1.67 MW DRER unit at Bus 61. The improvement in voltages of lowest voltage bus is also noticed here.

Without the installation of DRER unit in IEEE 69 bus network, the maximum voltage deviation of 0.099 p.u. is observed in constant load model, whereas, after the addition of the DRER unit at Bus. 61, this value is reduced to 0.024 p.u. The weakest buses in IEEE 69 bus test network without and with the installation of DRER unit are found Bus 65 and Bus 27, with the voltage magnitudes of 0.909 p.u. and 0. 967 p.u., respectively.

The simulation results of the each voltage dependent load models having two DRER units in IEEE 33 bus and IEEE 69 bus test cases are provided in Tables 4.4 and 4.5.

With the addition of two DRER units in both the test cases, the total size of DRER (sum of DRER Unit 1 & Unit 2) is significantly greater than the size of DRER for each voltage dependent load model in previous case (with 1 DRER unit only). With the increments in the penetration of DRER power, the reduction in total power losses is also observed to be much higher as compared to the reduction of total power losses with 1 DRER unit only. For the IEEE 33 bus test case, the optimum locations of DRER units for each load model are Bus 14 and Bus 30. Also, for the IEEE 69 bus test case, the optimum locations of DRER units for each load model are Bus 20 and Bus 61.

For the IEEE 33 bus test case, the minimum reduction in total power loss is seen in the constant load model in which 58.60% of total energy losses are reduced with the installations of 0.8 MW and 1.2 MW DRER units at Bus 14 and Bus 30, respectively. Also, the maximum reduction in total power losses is seen in industrial load model, in which 69.87% of the total power losses are reduced with the installation of 0.79 and 1.16 MW DRER units at Bus 13 and Bus 30, respectively. The addition of second DRER unit in IEEE 33 bus network also improved the bus voltages profile. After the installation of two DRER units, the maximum voltage deviation of 0.017 p.u is observed in constant load model.

The addition of second DRER unit in IEEE 33 bus test case enhanced the voltage of weak busses and now minimum voltage for industrial, residential and commercial loads are found at Bus 25. The voltage magnitude of Bus 25 after the

**Table 4.4** Summary of results with two DRER units in IEEE 33 bus test system

| Parameters | Constant | | Industrial | | Residential | | Commercial | |
|---|---|---|---|---|---|---|---|---|
| | Without | With DRER | Without | With DRER | Without | With DRER | Without | With DRER |
| DRER size (MW)@Bus no | | 1.2@30<br>0.8@14 | | 0.79@14<br>1.16@30 | | 0.78@14<br>1.18@30 | | 0.76@14<br>1.18@30 |
| Total power losses (kW) | 211.00 | 87.36 | 163.69 | 49.31 | 159.12 | 56.85 | 152.63 | 59.66 |
| Reduction in power loss (%) | | 58.60% | | 69.87% | | 64.27% | | 60.91% |
| Total power loss index | | 0.41 | | 0.30 | | 0.36 | | 0.39 |
| Voltage deviations | 0.134 | 0.017 | 0.100 | 0.008 | 0.099 | 0.008 | 0.095 | 0.007 |
| Min V@bus | 0.904@18 | 0.969@18 | 0.917@18 | 0.979@25 | 0.918@18 | 0.98@25 | 0.92@18 | 0.98@25 |

**Table 4.5** Summary of results with two DRER in IEEE 69 bus test system

| Parameters | Constant | | Industrial | | Residential | | Commercial | |
|---|---|---|---|---|---|---|---|---|
| | Without | With DRER | Without | With DRER | Without | With DRER | Without | With DRER |
| DRER size (MW)@Bus no | | 0.32@20 1.93@61 | | 0.4@20 1.84@61 | | 1.82@61 0.39@20 | | 0.38@20 1.8@61 |
| Total power losses (kW) | 224.95 | 73.85 | 171.42 | 31.67 | 164.91 | 41.35 | 156.97 | 45.61 |
| Reduction in power loss (%) | | 67.17% | | 81.52% | | 74.92% | | 70.94% |
| Total power loss index | | 0.33 | | 0.18 | | 0.25 | | 0.29 |
| Voltage deviations | 0.099 | 0.008 | 0.080 | 0.004 | 0.076 | 0.003 | 0.072 | 0.003 |
| Min V@bus | 0.909@65 | 0.983@65 | 0.92@65 | 0.989@69 | 0.922@65 | 0.989@27 | 0.924@65 | 0.989@27 |

installation of two DRER units is 0.979, 0.98, and 0.98 p.u., respectively. After the installation of two DRER units, now the weakest bus for constant load model is found Bus 33 with 0.969 p.u. voltage magnitude.

Similar to IEEE 33 bus test, the addition of second DRER unit in IEEE 69 bus test case also showed positive impact on the voltage magnitude of the weakest bus. Moreover, the minimum reduction in total power losses for this distribution network is observed for constant load model, in which 67.17% of total power losses are reduced by adding 0.32 and 1.93 MW DRER units at Bus 20 and Bus 61, respectively. The maximum reduction in total power losses is seen in industrial load model, in which more than 81.52% of the total power losses are reduced by adding 0.4 and 1.84 MW DRER units at Bus 20 and Bus 61, respectively. The addition of second DRER unit in IEEE 69 bus network also reduced the maximum voltage deviation of constant load model to 0.008 p.u., which was 0.099 p.u. in the case of single DRER unit.

The significance of installing the third DRER unit for the reducing the power losses in both test cases is also revealed. However, the addition of third DRER unit could not bring further improvement to the bus voltage. This is mainly due the limitations of penetration levels. The total size of the three DRER units is almost same as the total size of two DRER units. The simulation results for three DRER units in IEEE 33 bus and IEEE 69 bus test cases are provided in Tables 4.6 and 4.7. For each load model in IEEE 33 bus test case, the optimum locations of the DRER units are Bus 14, Bus 25 and Bus 31. Also, the locations of DRER units for different lode models in IEEE 69 bus test case are Bus 17, Bus 61 and Bus 64.

The reduction in total power losses for each load model in IEEE 33 bus network due to installation of third DRER was also experienced. Constant load model was the least beneficial among the four load models and only 62.94% of total power losses were reduced. The sizes of DRER were 0.89 MW at Bus 14, 0.45 MW at Bus 25 and 0.96 MW at Bus 31.

Similarly, industrial load model was the most beneficial among the four load models in IEEE 33 bus network and 76.07% of total power losses were reduced. The sizes of DRER for industrial load model were 0.82 MW at Bus 14, 0.54 MW at Bus 25 and 0.97 MW at Bus 30. The bus voltage profiles for the IEEE 69 bus network, with and without the installation of three DRER units are provided in Figs. 4.5, 4.6, 4.7 and 4.8.

The reduction in total power losses for each load model in IEEE 69 bus network due to installation of third DRER was also experienced. Constant load model was the least beneficial among the four load models and only 67.37% of total power losses were reduced. The sizes of DRER were 0.36 MW at Bus 17, 1.05 MW at Bus 61 and 0.84 MW at Bus 64.

Similarly, industrial load model was the most beneficial among the four load models and 82.22% of total power losses were reduced. The sizes of DRER for industrial load model were 0.47 MW at Bus 17, 1.39 MW at Bus 61 and 0.38 MW at Bus 64. The bus voltage profiles for the IEEE 69 bus network, with and without the installation of three DRER units are provided in Figs. 4.9, 4.10, 4.11, and 4.12.

**Table 4.6** Summary of results with three DRER units in IEEE 33 bus test system

| Parameters | Constant | | Industrial | | Residential | | Commercial | |
|---|---|---|---|---|---|---|---|---|
| | Without | With DRER | Without | With DRER | Without | With DRER | Without | With DRER |
| DRER size (MW)@Bus no | | 0.89@14<br>0.45@25<br>0.96@31 | | 0.97@31<br>0.82@14<br>0.54@25 | | 0.97@31<br>0.79@14<br>0.54@25 | | 0.78@14<br>0.96@31<br>0.54@25 |
| Total power losses (kW) | 211.00 | 78.20 | 163.69 | 39.18 | 159.12 | 46.88 | 152.63 | 49.77 |
| Reduction in power loss (%) | | 62.94% | | 76.07% | | 70.54% | | 67.39% |
| Total power loss index | | 0.37 | | 0.24 | | 0.29 | | 0.33 |
| Voltage deviations | 0.134 | 0.016 | 0.100 | 0.007 | 0.099 | 0.008 | 0.095 | 0.007 |
| Min V@bus | 0.904@18 | 0.969@30 | 0.917@18 | 0.98@8 | 0.918@18 | 0.979@29 | 0.92@18 | 0.98@29 |

**Table 4.7** Summary of results with three DRER units in IEEE 69 bus test system

| Parameters | Constant | | Industrial | | Residential | | Commercial | |
|---|---|---|---|---|---|---|---|---|
| | Without | With DRER | Without | With DRER | Without | With DRER | Without | With DRER |
| DRER size (MW)@Bus no | | 1.05@61 0.84@64 0.36@17 | | 1.39@61 0.38@64 0.47@17 | | 0.43@64 1.32@61 0.41@17 | | 0.53@64 1.26@61 0.44@17 |
| Total power losses (kW) | 224.95 | 73.40 | 171.42 | 30.48 | 164.91 | 40.32 | 156.97 | 45.79 |
| Reduction in power loss (%) | | 67.37% | | 82.22% | | 75.55% | | 70.83% |
| Total power loss index | | 0.33 | | 0.18 | | 0.24 | | 0.29 |
| Voltage deviations | 0.099 | 0.008 | 0.080 | 0.003 | 0.076 | 0.003 | 0.072 | 0.003 |
| Min V@bus | 0.909@65 | 0.982@27 | 0.92@65 | 0.99@69 | 0.922@65 | 0.989@69 | 0.924@65 | 0.99@27 |

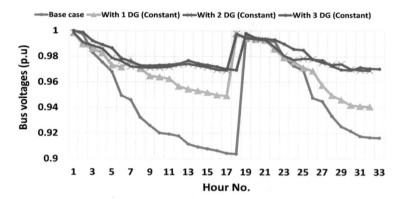

**Fig. 4.5** Bus voltage profile with and without DRER for constant load in IEEE 33 bus network

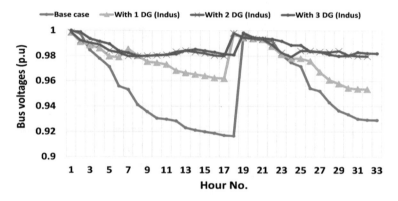

**Fig. 4.6** Bus voltage profile with and without DRER for industrial load in IEEE 33 bus network

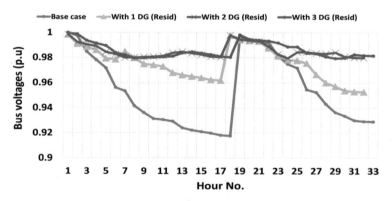

**Fig. 4.7** Bus voltage profile with and without DRER for residential load in IEEE 33 bus network

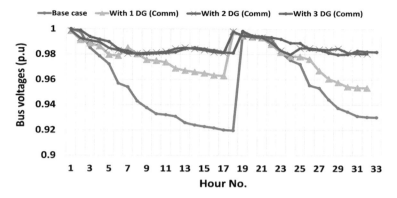

**Fig. 4.8** Bus voltage profile with and without DRER for commercial load in IEEE 33 bus network

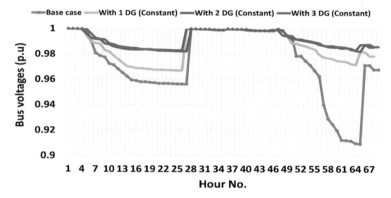

**Fig. 4.9** Bus voltage profile with and without DRER for constant load in IEEE 69 bus network

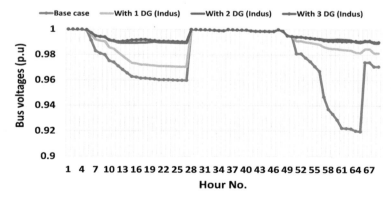

**Fig. 4.10** Bus voltage profile with and without DRER for constant load in IEEE 33 bus network

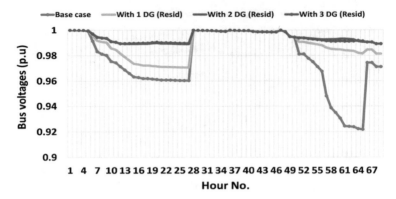

**Fig. 4.11** Bus voltage profile with and without DRER for residential load in IEEE 69 bus network

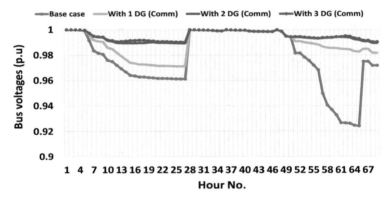

**Fig. 4.12** Bus voltage profile with and without DRER for commercial load in IEEE 69 bus network

## 4.5 Summary

The chapter presents a weighted sum approach by Mixed Integer Optimization with Genetic Algorithm (MIOGA) for the planning of Distributed Energy Resources (DRER) in the distribution networks. The study considered solar photovoltaic (PV) technology for as a source of active power. The objective functions used during the optimization were Total Power Loss Index (TPLI), Voltage Deviations (VDs) and minimization of the total size of DRER units. To test the effectiveness of the proposed algorithm, different voltage dependent load models were adopted. The proposed algorithm was applied on IEEE 33 bus and IEEE 69 bus test distribution networks. From the study, the followings can be concluded:

1. For different voltage dependent loads, the reduction in total power losses with the installation of single DRER unit in IEEE 33 and IEEE 69 test distribution network were ranging between 43.68–52.46% and 62.35–75.24%, respectively.
2. Compared to single DRER unit, the installation of multiple DRER units in the distribution networks significantly has reduced power losses and at the same time enhanced the penetration of DRER power. The reduction in total power losses with the installation of two DRER units in IEEE 33 and IEEE 69 test distribution network were ranging in between 58.60–69.87% and 67.17–81.52%, respectively. The reduction in total power losses with the installation of three DRER units in IEEE 33 and IEEE 69 test distribution network were ranging between 62.94–76.07% and 67.37–82.22%, respectively.
3. The installation of multiple DRER units has also provided positive impact on bus voltage profiles in both test distribution networks. The enhancement in voltage deviations due to addition of two DRER units and three DRER units was found same in both distribution networks. This mainly due to total size of two DRER units and total size of three DRER units were almost same.
4. The proposed algorithm was seen effective for solving the multi-objective optimization problem for the planning of Distributed Energy Resources (DRER) in the distribution networks.

# References

1. Y.H. Liu, Z.Q. Wu, Y.Q. Tu, Q.Y. Huang, H.W. Lou, A survey on distributed generation and its networking technology. Power Syst. Technol. **15**, 020 (2008)
2. C. Kost, J.N. Mayer, J. Thomsen, N. Hartmann, C. Senkpiel, S. Philipps et al., *Levelized Cost of Electricity Renewable Energy Technologies*.(1 March 2013). Available: https://www.ise.fraunhofer.de/content/dam/ise/en/documents/publications/studies/Fraunhofer-ISE_LCOE_Renewable_Energy_technologies.pdf
3. A. Lee, O. Zinaman, J. Logan, M. Bazilian, D. Arent, R.L. Newmark, Interactions, complementarities and tensions at the nexus of natural gas and renewable energy. Electr. J. **25**, 38–48 (2012)
4. D.Q. Hung, N. Mithulananthan, R.C. Bansal, Analytical expressions for DG allocation in primary distribution networks. IEEE Trans. Energ. Convers. **25**, 814–820 (2010)
5. N. Acharya, P. Mahat, N. Mithulananthan, An analytical approach for DG allocation in primary distribution network. Int. J. Electr. Power Energ. Syst. **28**, 669–678 (2006)
6. P.S. Georgilakis, N.D. Hatziargyriou, Optimal distributed generation placement in power distribution networks: models, methods, and future research. IEEE Trans. Power Syst. **28**, 3420–3428 (2013)
7. W.-S. Tan, M.Y. Hassan, M.S. Majid, H.A. Rahman, Optimal distributed renewable generation planning: a review of different approaches. Renew. Sustain. Energ. Rev. **18**, 626–645 (2013)
8. M. Kowsalya, Optimal size and siting of multiple distributed generators in distribution system using bacterial foraging optimization. Swarm Evol. Comput. **15**, 58–65 (2014)
9. M.H. Moradi, M. Abedini, A combination of genetic algorithm and particle swarm optimization for optimal DG location and sizing in distribution systems. Int. J. Electr. Power Energ. Syst. **34**, 66–74 (2012)

10. S.K. Injeti, N.P. Kumar, A novel approach to identify optimal access point and capacity of multiple DGs in a small, medium and large scale radial distribution systems. Int. J. Electr. Power Energ. Syst. **45**, 142–151 (2013)
11. D.R. Prabha, T. Jayabarathi, Optimal placement and sizing of multiple distributed generating units in distribution networks by invasive weed optimization algorithm. Ain Shams Eng. J. **7**, 683–694 (2016)
12. H. Manafi, N. Ghadimi, M. Ojaroudi, P. Farhadi, Optimal placement of distributed generations in radial distribution systems using various PSO and DE algorithms. Elektron. Elektrotech. **19**, 53–57 (2013)
13. B. Mohanty, S. Tripathy, A teaching learning based optimization technique for optimal location and size of DG in distribution network. J. Electr. Syst. Inform. Technol. **3**, 33–44 (2016)
14. R.S. Rao, K. Ravindra, K. Satish, S. Narasimham, Power loss minimization in distribution system using network reconfiguration in the presence of distributed generation. IEEE Trans. Power Syst. **28**, 317–325 (2013)
15. M. Jamil, A.S. Anees, Optimal sizing and location of SPV (solar photovoltaic) based MLDG (multiple location distributed generator) in distribution system for loss reduction, voltage profile improvement with economical benefits. Energy **103**, 231–239 (2016)
16. C. Yammani, S. Maheswarapu, S.K. Matam, A multi-objective Shuffled Bat algorithm for optimal placement and sizing of multi distributed generations with different load models. Int. J. Electr. Power Energ. Syst. **79**, 120–131 (2016)
17. H. Karami, B. Zaker, B. Vahidi, G.B. Gharehpetian, Optimal multi-objective number, locating, and sizing of distributed generations and distributed static compensators considering loadability using the genetic algorithm. Electr Power Compon. Syst. **44**, 2161–2171 (2016)
18. M. Esmaeili, M. Sedighizadeh, M. Esmaili, Multi-objective optimal reconfiguration and DG (Distributed generation) power allocation in distribution networks using Big Bang-Big Crunch algorithm considering load uncertainty. Energy **103**, 86–99 (2016)
19. M.H. Moradi, S.R. Tousi, M. Abedini, Multi-objective PFDE algorithm for solving the optimal siting and sizing problem of multiple DG sources. Int. J. Electr. Power Energ. Syst. **56**, 117–126 (2014)
20. K. Mahesh, P. Nallagownden, I. Elamvazuthi, Advanced pareto front non-dominated sorting multi-objective particle swarm optimization for optimal placement and sizing of distributed generation. Energies **9**, 982 (2016)
21. M.A. Darfoun, M.E. El-Hawary, Multi-objective optimization approach for optimal distributed generation sizing and placement. Electr. Power Compon. Syst. **43**, 828–836 (2015)
22. N. Mohandas, R. Balamurugan, L. Lakshminarasimman, Optimal location and sizing of real power DG units to improve the voltage stability in the distribution system using ABC algorithm united with chaos. Int. J. Electr. Power Energ. Syst. **66**, 41–52 (2015)
23. W. Jakob, C. Blume, Pareto optimization or cascaded weighted sum: a comparison of concepts. Algorithms **7**, 166 (2014)
24. R.T. Marler, J.S. Arora, The weighted sum method for multi-objective optimization: new insights. Struct. Multi. Optim. **41**, 853–862 (2010)
25. A. Konak, D.W. Coit, A.E. Smith, Multi-objective optimization using genetic algorithms: a tutorial. Reliab. Eng. Syst. Saf. **91**, 992–1007 (2006)
26. G. Celli, E. Ghiani, S. Mocci, F. Pilo, A multiobjective evolutionary algorithm for the sizing and siting of distributed generation. IEEE Trans. Power Syst. **20**, 750–757 (2005)
27. M.A. Abido, Environmental/economic power dispatch using multiobjective evolutionary algorithms. IEEE Trans. Power Syst. **18**, 1529–1537 (2003)
28. A. Alarcon-Rodriguez, G. Ault, S. Galloway, Multi-objective planning of distributed energy resources: a review of the state-of-the-art. Renew. Sustain. Energ. Rev. **14**, 1353–1366 (2010)
29. K. Deb, *Multi-Objective Optimization Using Evolutionary Algorithms*, vol 16. (Wiley, 2001)
30. M.C. Neri, A.Y. Hermosilla, Decomposition-Metaheuristic method applied to a capacitated facility location problem, in *Proceedings of the Third Asian Mathematical Conference 2000*, University of the Philippines, Diliman, Philippines, 23–27 Oct 2000, 2002, p. 389

31. D.E. Goldberg, J.H. Holland, Genetic algorithms and machine learning. Mach. Learn. **3**, 95–99 (1988)
32. M. Mitchell, *An Introduction to Genetic Algorithms* (MIT press, 1998)
33. K.-H. Kim, Y.-J. Lee, S.-B. Rhee, S.-K. Lee, and S.-K. You, Dispersed generator placement using fuzzy-GA in distribution systems, in *Power Engineering Society Summer Meeting*, 2002 IEEE, pp. 1148–1153
34. M.H. Moradi, A. Zeinalzadeh, Y. Mohammadi, M. Abedini, An efficient hybrid method for solving the optimal sitting and sizing problem of DG and shunt capacitor banks simultaneously based on imperialist competitive algorithm and genetic algorithm. Int. J. Electr. Power Energ. Syst. **54**, 101–111 (2014)
35. M. Gandomkar, M. Vakilian, M. Ehsan, A combination of genetic algorithm and simulated annealing for optimal DG allocation in distribution networks, in *Canadian Conference on Electrical and Computer Engineering*, 2005, pp. 645–648
36. T. Yokota, M. Gen, Y.-X. Li, Genetic algorithm for non-linear mixed integer programming problems and its applications. Comput. Ind. Eng. **30**, 905–917 (1996)
37. J.A. Momoh, M. El-Hawary, R. Adapa, A review of selected optimal power flow literature to 1993. II. Newton, linear programming and interior point methods. IEEE Trans. Power Syst. **14**, 105–111 (1999)
38. A. Dukpa, B. Venkatesh, M. El-Hawary, Application of continuation power flow method in radial distribution systems. Electr. Power Syst. Res. **79**, 1503–1510 (2009)
39. D. Shirmohammadi, H.W. Hong, A. Semlyen, G. Luo, A compensation-based power flow method for weakly meshed distribution and transmission networks. IEEE Trans. Power Syst. **3**, 753–762 (1988)
40. A. Tah, D. Das, Novel analytical method for the placement and sizing of distributed generation unit on distribution networks with and without considering P and PQV buses. Int. J. Electr. Power Energ. Syst. **78**, 401–413 (2016)
41. S. Bolognani, S. Zampieri, On the existence and linear approximation of the power flow solution in power distribution networks. IEEE Trans. Power Syst. **31**, 163–172 (2016)
42. M. Haque, Efficient load flow method for distribution systems with radial or mesh configuration. IEE Proc-Gener. Transm. Distrib. **143**, 33–38 (1996)
43. J.-H. Teng, A direct approach for distribution system load flow solutions. IEEE Trans. Power Deliv **18**, 882–887 (2003)
44. S. Mishra, A simple algorithm for unbalanced radial distribution system load flow, in *TENCON 2008–2008 IEEE Region 10 Conference*, 2008, pp. 1–6
45. G. Chang, S. Chu, H. Wang, An improved backward/forward sweep load flow algorithm for radial distribution systems. IEEE Trans. Power Syst. **22**, 882–884 (2007)
46. E. Bompard, E. Carpaneto, G. Chicco, R. Napoli, Convergence of the backward/forward sweep method for the load-flow analysis of radial distribution systems. Int. J. Electr. Power Energ. Syst. **22**, 521–530 (2000)
47. V. Murty, B.R. Teja, A. Kumar, A contribution to load flow in radial distribution system and comparison of different load flow methods, in *2014 International Conference on Power Signals Control and Computations (EPSCICON)*, 2014, pp. 1–6
48. T. Alinjak, I. Pavić, M. Stojkov, Improvement of backward/forward sweep power flow method by using modified breadth-first search strategy. IET Gener. Transm. Distrib. **11**, 102–109 (2017)
49. D.Q. Hung, N. Mithulananthan, K.Y. Lee, Determining PV penetration for distribution systems with time-varying load models. IEEE Trans. Power Syst. **29**, 3048–3057 (2014)
50. L.A. Gallego, E. Carreno, and A. Padilha-Feltrin, Distributed generation modelling for unbalanced three-phase power flow calculations in smart grids, in *2010 IEEE/PES Transmission and Distribution Conference and Exposition: Latin America (T&D-LA)*, 2010, pp. 323–328
51. D. Singh, R. Misra, Effect of load models in distributed generation planning. IEEE Trans. Power Syst. **22**, 2204–2212 (2007)

52. D. Singh, K. Verma, Multiobjective optimization for DG planning with load models. IEEE Trans. Power Syst. **24**, 427–436 (2009)
53. D.Q. Hung, N. Mithulananthan, R. Bansal, Analytical strategies for renewable distributed generation integration considering energy loss minimization. Appl. Energ. **105**, 75–85 (2013)
54. TNB, Tenaga Nasional Berhad Annual Report 2015
55. E. Mahboubi-Moghaddam, M.R. Narimani, M.H. Khooban, A. Azizivahed, M. Javid sharifi, Multi-objective Distribution feeder reconfiguration to improve transient stability, and minimize power loss and operation cost using an enhanced evolutionary algorithm at the presence of distributed generations. Int. J. Electr. Power Energ. Syst. **76**, 35–43, 3 (2016)
56. D. Whitley, A genetic algorithm tutorial. Stat Comput **4**, 65–85 (1994)
57. J.H. Holland, Hierarchical descriptions, universal spaces and adaptive systems, Dtic Document (1968)
58. Z. Boor, S.M. Hosseini, Optimal placement of DG to improve the reliability of distribution systems considering time varying loads using genetic algorithm. Majlesi J. Electr. Eng vol. 7 (2012)
59. S. Naik, D. Khatod, M. Sharma, Sizing and siting of distributed generation in distribution networks for real power loss minimization using analytical approach, in *2013 International Conference on Power, Energy and Control (ICPEC)*, 2013, pp. 740–745
60. R. Chang, N. Mithulananthan, T. Saha, Novel mixed-integer method to optimize distributed generation mix in primary distribution systems, in *2011 21st Australasian Universities Power Engineering Conference (AUPEC)*, 2011, pp. 1–6
61. D.Q. Hung, N. Mithulananthan, An optimal operating strategy of DG unit for power loss reduction in distribution systems, in *2012 7th IEEE International Conference on Industrial and Information Systems (ICIIS)*, 2012, pp. 1–6
62. B. Venkatesh, R. Ranjan, H. Gooi, Optimal reconfiguration of radial distribution systems to maximize loadability. IEEE Trans. Power Syst. **19**, 260–266 (2004)

# Chapter 5
# Feasibility Study on Hybrid Renewable Energy to Supply Unmanned Offshore Platform

**Mohd Faris Abdullah, Moustafa Ahmed and Khairul Nisak Md Hasan**

Hybrid renewable energy system is introduced in this new era as an effective solution to provide remote residential villages, off-grid small commercial or industrial areas and off-grid offshore installations which do not require extensive generation capability with clean and cheap power supply. A hybrid system is usually composed of two or three different conventional and/or unconventional power sources such as diesel engine, solar, wind energy, etc. In Malaysia, oil and gas industry has proven to be one of the most successful and emerging industries and the main pillar for Malaysia economic growth during the last twenty-five years. The exploration and drilling operations have extended not only on the onshore scope but also on the offshore scope either in Malaysia or overseas venture. In this chapter, investigation on the potential utilization of hybrid renewable energy system in the Malaysian offshore environment is examined. The solar and wind energy potentials in Malaysia are investigated and the feasibility study of supplying unmanned offshore platform is studied by using meteorological data from the existing project location at South China Sea.

## 5.1 Introduction

Malaysia has been in the middle of a rapid economic growth since 1980 and is expected to reach its peak in 2020 in becoming a fully developed country. Oil and gas industry in Malaysia plays an important role in the country's economy. Throughout the past century, the country mainly relied on four main energy sources (oil, natural gas, hydropower and coal), with an increasing dependence on oil and natural gas reserves [1].

M. F. Abdullah (✉) · M. Ahmed · K. N. M. Hasan
Universiti Teknologi PETRONAS, Seri Iskandar, Malaysia
e-mail: mfaris_abdullah@utp.edu.my

© The Author(s), under exclusive license to Springer Nature Singapore Pte Ltd. 2018
S. A. Sulaiman et al. (eds.), *Sustainable Electrical Power Resources through Energy Optimization and Future Engineering*, SpringerBriefs in Energy,
https://doi.org/10.1007/978-981-13-0435-4_5

The majority of Malaysia recoverable hydrocarbon resources are located in the offshore part of the country. Offshore oil field development requires special consideration during the exploration, appraisal, field development, project implementation and field production phases. The scope of work and capital expenditure is generally greater for an offshore field development [2].

The development cost for offshore oil fields using manned installations require a capital investment of millions of dollars. Moreover, the operation and maintenance costs will be much higher due to the involvement of human factor. By comparison, the use of unmanned offshore platform would save a lot of resources, capital investments, and will reduce the operation and maintenance costs [3].

To provide sufficient power supply to the unmanned offshore platforms, the older generation platforms mainly used conventional power sources. The conventional power sources such as diesel and natural gas turbines are used to supply the instrumentation and electrical devices with continuous power supply [4].

## 5.2   Hybrid Renewable Energy Systems

Throughout the last two decades, the global awareness towards environmental issues has grown rapidly. Using fossil fuels as the main energy resource in power generation does not only cause a destructive environmental effect but also an increasing consumption of the world conventional energy sources. Proposing unconventional power sources such as solar energy (photovoltaic (PV) power generation) and wind energy (wind turbine power generation) as a solution for the power demand initiated a global movement towards the renewable energy [5].

### 5.2.1   Solar PV

The photoelectric effect depends on the P-N junction and the diffusion current resulted from the effect of solar radiation on P-N junction in semiconductors. Basically, in no-light mode, solar panels operate in the reverse direction and hence a drift current will be initiated between the P-N junctions. When solar panels are exposed to light, cell photons with energy greater than the P-N junction gap energy are absorbed and hence electron hole pairs are formed. These carriers are separated under the influence of electric fields within the junction, creating a current that is proportional to the incidence of solar irradiation [6].

Solar cells usually display nonlinear characteristics between generated power and cell voltage P-V curve and also between generated current and cell voltage I-V curve. Basically, a solar cell has fundamental parameters which form the solar cell characteristics and behavior upon exposing to light or radiation. These parameters are short circuit current ($I_{sc}$), open circuit voltage ($V_{oc}$), maximum power point tracking algorithm (MPPT) and efficiency of the solar cell ($\eta$) [7].

Accurate simulation can be obtained after considering the following parameters [8]:

- Temperature dependence of the diode reversed saturation current ($I_s$)
- Temperature dependence of the photo current ($I_{ph}$)
- Series resistance ($R_s$)
- Shunt resistance ($R_{sh}$).

Equations that define the model of a PV cell are:

$$V_T = \frac{K * T_{OP}}{q}$$ (5.1)

$$V_{OC} = V_T \ln \frac{I_{Ph}}{I_S}$$ (5.2)

$$I_d = [e^{\left(\frac{V + I*R_S}{n*V_t*C*N_S}\right)} - 1] * I_S * N_P$$ (5.3)

$$I_S = I_{rs} * \left(\frac{T_{op}}{T_{ref}}\right)^3 * e^{\left[\frac{q*E_g}{n*k}\left(\frac{1}{T_{op}} - \frac{1}{T_{ref}}\right)\right]}$$ (5.4)

$$I_{rs} = \frac{I_{sc}}{\left[e^{\left[\frac{q*V_{oc}}{n*k*C*T_{op}}\right]} - 1\right]}$$ (5.5)

$$I_{sh} = \frac{V + I * R_s}{R_P}$$ (5.6)

$$I_{ph} = G_k \left[I_{sc} + K_1(T_{OP} - T_{ref})\right]$$ (5.7)

$$I = I_{ph} * N_P - I_d - I_{sh}$$ (5.8)

where,

| | |
|---|---|
| STC | Standard Test Condition, $G$ = 1 kW/m$^2$, $T_{OP}$ = 25 °C |
| $G_k$ | Solar irradiance ratio |
| $V_T$ | Thermal voltage, V |
| $K$ | Boltzmann's constant, $1.38e^{-23}$ |
| $T_{OP}$ | Cell operating temperature in °C |
| $T_{ref}$ | Cell temperature at 25 °C |
| $q$ | Electron charge constant, $1.6e^{-19}$ C |
| $I_S$ | Diode reversed saturation current, A |
| $I_{rs}$ | Diode reversed saturation current at $T_{OP}$ |
| $I$ | Output current from the PV panel, A |
| $I_{sh}$ | Shunt current, A |

| $V$ | Output voltage from the PV panel, V |
|---|---|
| $n$ | Diode ideality factor, 1.36 |
| $C$ | Number of cells in a PV panel, 36 |
| $N_S, N_P$ | No of PV panel in series & parallel |
| $E_g$ | Band-gap energy of the cell, 1.12 eV. |

### 5.2.2 Wind Turbine

Wind turbines are classified into vertical and horizontal axis wind turbines. Vertical axis wind turbines do not require any towers and the generator, gear boxes and turbine control system can be installed on the ground. Moreover, controlling the amplitude and the frequency of a vertical axis wind turbine will be difficult and hence the power efficiency will be affected significantly. Currently, horizontal axis wind turbines are the most affordable type of wind turbines available in the market [9].

The following equations are used to model the blades mechanism and to identify the mathematical relationship between the wind speed and the generated mechanical power:

$$E = 0.5 * m * v^2 \tag{5.9}$$

$$P_W = \frac{dE}{dt} = 0.5 * m * v^2 \tag{5.10}$$

$$P_W = 0.5 * \rho * A * v^3 \tag{5.11}$$

$$\rho = \rho_0 - 1.194 * 10^{-14} * H \tag{5.12}$$

$$\rho_0 = 1.225 \ \frac{kg}{m^3} \ at \ T = 298 \ K \tag{5.13}$$

$$P_{Blade} = C_{P(\lambda,\beta)} * P_W = C_{P(\lambda,\beta)} * 0.5 * m * v^3 \tag{5.14}$$

$$\lambda = \frac{\omega_m * R}{v} \tag{5.15}$$

$$T_w = \frac{P_{Blade}}{\omega_m} = C_{P(\lambda,\beta)} * 0.5 * \rho * A * v^3 \tag{5.16}$$

where,

| $E$ | Kinetic energy in air |
|---|---|
| $m$ | Mass flow rate (kg/sec) |
| $v$ | Wind speed (m/s$^2$) |

$P_W$     Power in the moving air (W)
$P$       Air density (kg/m$^3$)
$\rho_0$  Air density at sea level at temperature $T = 298$ K, $\rho_0 = 1.225$ kg/m$^3$
$H$       Height above sea level (m)
$P_{Blade}$ Power extracted from the wind (W)
$\lambda$ Tip speed ratio (unit less)
$\omega_m$ Angular velocity of the rotor (rad/s)
$R$       Rotor radius (blade length) in meter
$T_w$     Rotor torque to the generator (Nm).

## 5.2.3  Diesel Generator

Diesel generating system has been used to supply electricity for remote areas, offshore constructions and off-grid societies. The system has always showed robust and consistent performance [10]. The engine is used to produce mechanical energy which is transferred to the rotor of the synchronous machine via the rotating shaft between the two sets. Finally, the generator converts the mechanical energy to electrical energy [11].

## 5.2.4  Energy Storage System (Battery System)

Due to the irregularity and fragmentary of the renewable energy sources, which are caused by variations in solar irradiation levels as well as wind speeds, it is highly doubtful that the AC/DC loads on board of the platforms will be only supplied by the renewable sources. A battery system needs to be implemented in the hybrid renewable energy system along with the renewable energy units as well as the emergency diesel generator.

## 5.3  Solar and Wind Energy Potential in Malaysia

Meteorological data such as solar radiation, wind speed and temperature profiles differ from a location to another. To effectively analyze the potential of solar and wind energy for the project location, the meteorological data for the location should be investigated to determine the energy profile of an area.

### 5.3.1   Solar Energy Potential

According to the world solar radiation distribution, the solar radiation levels in Malaysia are considered amongst the moderate in the world. Malaysia has a mean solar radiation of 4.0–4.9 kWh/m$^2$/day while the mean solar radiation level for the most countries in the world is around 6.0–6.9 kWh/m$^2$/day [12, 13].

### 5.3.2   Wind Energy Potential

Wind speed, wind flow dynamics and climatic conditions are among the most crucial factors that designers rely on during pre-feasibility study. In Malaysia, the meteorological data needed for the study can be obtained and sourced from the national meteorological center [14, 15].

The main criteria for the selection was that during the annual northeast monsoon season, the wind speed in these areas would exceed 5 m/s. Despite the high wind speed during the monsoon, on average, the sites experienced low wind speed [16].

## 5.4   Unmanned Offshore Platform Load

Based on the experience of an existing North Sea unmanned offshore platform in Cutter area, the DC load data would vary between 250 and 400 W. The probability for higher load is considered as rare and would be unlikely to occur. The average daily demand of current on-board the platform for the instrumentation devices is 13.85 A for a DC voltage value of 24 V as specified by Shell, the owner company of the platform [17]. Consequently, the total average load for the unmanned off-shore platform is calculated via:

$$P_{watt} = I_{avg} V_{avg} \tag{5.17}$$

$$E_{kwh} = \frac{P_{watt} t_h}{1000} \tag{5.18}$$

Based on the Eqs. (5.17) and (5.18), the total average load will be 332.4 W and this will result in a total energy consumption of 7.978 kWh/day.

## 5.5   Optimal Study

The aim of the optimization study is to achieve an optimal design of the hybrid renewable energy system components with the lowest cost. The study is carried out by using HOMER software, which was earlier introduced by the American National

Renewable Energy Laboratory (NREL). Originally, HOMER was dedicated for optimization studies of off-grid hybrid renewable energy systems. Moreover, the study determines the ratings and sizing of each system components during the lifetime which is estimated to be 20 years (based on the lifespan of PV panels).

## 5.5.1  Proposed Optimization Technique

The proposed optimization technique is a parametric based approach. The optimization model takes inputs from the hybrid renewable energy system components as well as the wind data, solar data and diesel generator fuel cost. In addition to that, and seeking for the simplicity of the analysis, the power ratings of the hybrid renewable energy system components are fixed. However, the optimization model varies the number of units for each component, scaled annual irradiation level for the solar data and scaled annual wind speed for the wind data.

## 5.5.2  Cost Objective Function

The cost objective function purpose is to reduce the total cost of the project during its total lifetime. Here, lifetime is considered as the lifetime of the solar PV modules as they have the longest lifespan amongst all the system components. The cost objective function of the project is represented by the net present cost ($NPC$) investment starting from the installation and commissioning cost to a discounted value of operation and maintenance costs and hybrid renewable energy system components replacement cost. The inflation rate during the lifetime is also be taken into considerations.

The general equation for the cost objective function ($OF_{Cost}$), which is used in the optimization model includes capital cost ($CC$), annual operation and maintenance cost ($OMC$), battery banks replacement cost ($RC_{Cost-Batteries}$), replacement cost of the solar PV modules ($RC_{PV}$), replacement cost of the wind turbines ($RC_{WT}$) and the diesel generator fuel cost ($FC$). All the costs are calculated based on the total lifetime. However, the cost for inverters, rectifiers, system cables and the overall control system are not taken into considerations because they represent a small share of the total system cost as compared to those of the major system components. Consequently, the general equation of the cost objective function can be written as:

$$OF_{Cost} = CC + OMC + RC_{Cost-Batteries} + RC_{WT} + RC_{PV} + FC \qquad (5.19)$$

## 5.6   Results and Discussion

Since renewable energy depends on the natural resources, the locality of the proposed site is determined first in order to gather meteorological data. The simulation results of solar PV, wind turbine and diesel generator are presented before finally the optimal study is performed.

### 5.6.1   Project Location and Meteorological Data

The project location is selected to be 120 km from the shores of Sabah and Sarawak state and in line with the city of Bintulu. The coordination of the selected location is, longitude: 4.088 and latitude: 112.567.

The monthly average temperature profile data is obtained using the database from NASA Surface Meteorology and Solar Energy. The database is gathered and analyzed over 22 years by using several satellites for data collection and analysis. Figure 5.1 presents the maximum, minimum and average monthly temperature profiles.

The monthly average solar irradiation and daylight hours are presented in Fig. 5.2 to determine the potential of solar energy and the amount of power that can be generated by utilizing solar panels.

One of the most important key points of the feasibility study is to identify the wind profile and solar irradiation levels in the selected region. Moreover, analysis of both profiles determines the total electrical power production of the system. In Fig. 5.3, solar irradiation and wind speed profiles are presented. It can be shown that wind speed profile does not have a regular pattern just like the solar irradiance profile.

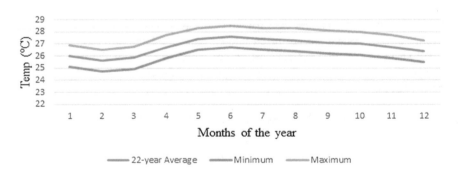

**Fig. 5.1**  Temperature profile at South China Sea

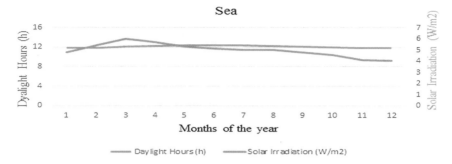

**Fig. 5.2** Hours of daylight-solar irradiation at South China Sea

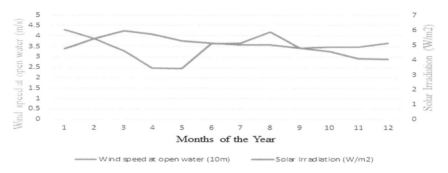

**Fig. 5.3** Solar irradiation and wind speed profiles at South China Sea

## 5.6.2 Solar PV

Simpowersystems block representation is used to implement the mathematical equations described in Sect. 5.2.1. The model schematic is shown in Fig. 5.4. The model is constructed for solar PV system of two parallel arrays with 10 series modules in each array. The maximum output power $P_{mpp}$ from the system is calculated to be 1100 W according to Solarex-MSX60 60Wp Solar Module characteristics used in the model. The model is tested in standard conditions, irradiation level of 1000 W/m$^2$ and operating temperature of 25 °C.

In Fig. 5.5, the final Input-Output values for the buck convertor are shown. The inputs for the convertor are the solar PV system (generation side) and the output values represent the output resistance (load side). The maximum output power produced by the PV array is maintained despite the changes in the load value. Also, the convertor steps down the output voltage from 170 V in the input side to 100 V on the load side and hence the output current increases from 6.5 A on the input side to 11.6 A on the output side.

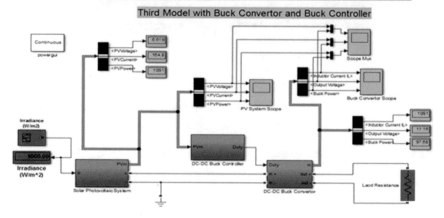

**Fig. 5.4** The PV model with buck convertor and controller

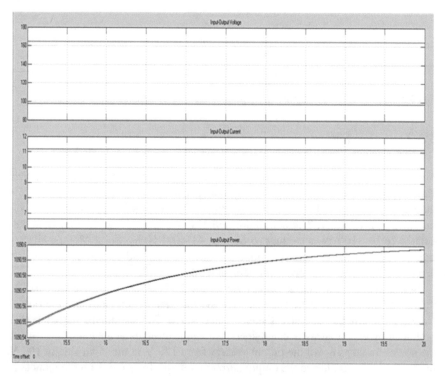

**Fig. 5.5** Input-output values for the buck convertor

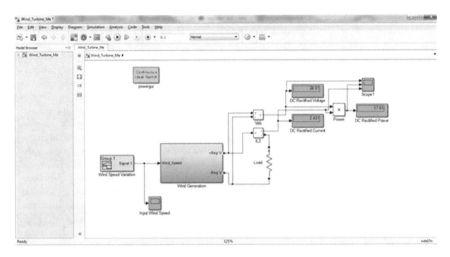

**Fig. 5.6** Wind turbine model

### 5.6.3 Wind Turbine

The simulated model using Simpowersystems consisted of Wind Turbine Rotor Blades Aeronautics, Two-mass Mechanical Drivetrain, Permanent Magnet Synchronous Generator, Universal Bridge (Rectifier Circuit) and Voltage Regulator Circuit blocks. The full model can be seen in Fig. 5.6. The output regulated values for voltage, current and power are concluded in Fig. 5.7.

### 5.6.4 Diesel Generator

The diesel generator modeling and simulation has been performed to analyze the performance of the diesel generator and the contribution of the excitation system, speed governor and automatic voltage regulator. The diesel generator model showed high stability and reliability while varying the output loads. The complete model of the diesel generator and the synchronous generator output voltage are shown in Figs. 5.8 and 5.9 respectively.

### 5.6.5 Optimal Study

The optimize parameters of the hybrid renewable energy system components obtained in Sects. 5.6.1–5.6.4 are used in HOMER software. Table 5.1 tabulates the solar PV parameters.

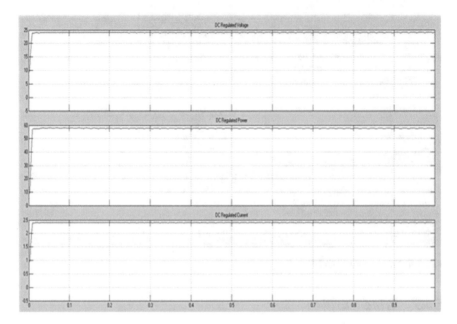

**Fig. 5.7** Regulated values of the wind turbine model

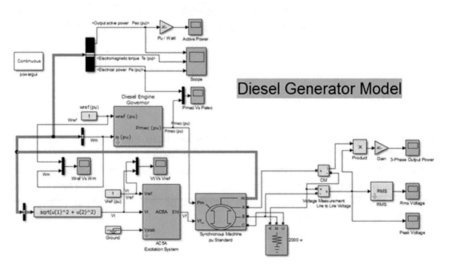

**Fig. 5.8** Diesel generator model

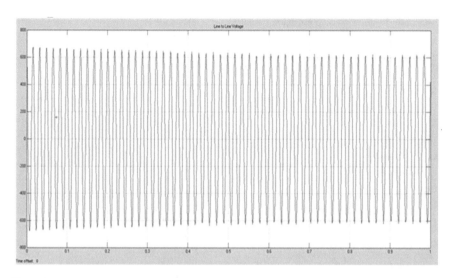

**Fig. 5.9** Synchronous generator output voltage

**Table 5.1** Solar PV optimization parameters

| Module rated power (W) | 300 |
|---|---|
| Module lifetime (years) | 20 |
| Module efficiency (%) | 13 |
| Module capital cost ($) | 500 |
| Module replacement cost ($) | 400 |
| Module OMC ($) | 0 |
| Number of modules for the simulation | 1–10 units |
| Nominal operating temperature (°C) | 25 |
| Fixed percentage of replacement (%) | 0.0112 |
| Average annual irradiation level (kWh/m$^2$/d) | 4.97 |
| Project latitude | 4.008 |
| Project longitude | 112.567 |

Table 5.2 shows the wind turbine parameters. The wind turbine used is a 1 kW horizontal axis Future Energy wind turbine.

The diesel generator parameters are shown in Table 5.3. The diesel generator is utilized in the project as a backup generation set and the renewable fraction of generation depends on the diesel fuel price as well as the metrological data.

In Table 5.4, the battery bank and AC/DC convertor parameters are listed. Battery bank used in the optimization model is Vision 6FM200D.

**Table 5.2** Wind turbine optimization parameters

| Manufacturer | Future Energy |
|---|---|
| Turbine model | AF1-24v-0125 (406 PMG) |
| Turbine Blades | 3 blades |
| Startup Wind Speed | 2 m/s |
| Charging Initiation Wind Speed | 3 m/s |
| Rated power (kW) | 1 kW DC |
| Lifetime (years) | 15 |
| Capital cost ($) | 3000 |
| Replacement cost ($) | 2500 |
| OMC ($) | 50 |
| Number of modules for the simulation | 1–3 units |
| Fixed percentage of replacement (%) | 0.0112 |
| Hub height (m) | 6 |
| Annual average wind speed (m/s) | 4.5 |
| Project latitude | 4.008 |
| Project longitude | 112.567 |

**Table 5.3** Diesel generator optimization parameters

| Rated power (kW) | 1 kW AC |
|---|---|
| Lifetime (operating hours) | 15,000 |
| Capital cost ($) | 1500 |
| Replacement cost ($) | 1200 |
| OMC ($/hr) | 0.05 |
| Number of modules for the simulation | 1–2 units |
| Fixed percentage of replacement (%) | 0.0112 |
| Maximum load ratio (%) | 30 |
| Generator fuel | Diesel |
| Carbon monoxide emission factor (g/l) | 6.5 |
| Diesel Price ($/L) | 0.35–0.6 |

The optimization model results are shown in Fig. 5.10 where the complete hybrid renewable energy system schematic and the optimal system components are presented.

In Fig. 5.11, a detailed look on the proposed optimal system components are presented. It can be seen that there are four proposed designs for the system which fulfill the cost objective function. However, the first proposed design is the most optimum design for the hybrid renewable energy system in terms of cost reduction. Despite the increase in the renewable fraction of the second design to 60%, the first design is proved to be the lowest cost with a renewable fraction of 45%.

**Table 5.4** Battery and convertor optimization parameters

| | |
|---|---|
| Battery nominal voltage (V) | 12 |
| Battery nominal capacity (Ah) | 200 Ah (2.4 kWh) |
| Battery lifetime throughput (kWh) | 917 |
| Battery capital cost ($) | 250 |
| Battery replacement cost ($) | 200 |
| Battery OMC ($) | 20 |
| Number of battery modules for the optimization | 1–12 units |
| Fixed percentage of replacement (%) | 1 |
| Convertor rated power (kW) | 1 |
| Number of convertor modules for the optimization | 1–3 units |
| Convertor capital cost ($) | 1000 |
| Convertor replacement cost ($) | 1000 |
| Convertor OMC ($) | 100 |
| Convertor lifetime (years) | 15 |
| Convertor efficiency (%) | 85–90 |

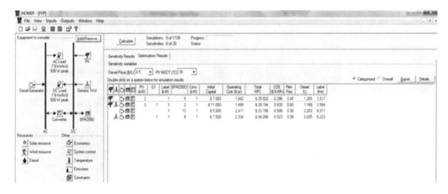

**Fig. 5.10** Optimization model results

| | | PV (kW) | G1 (kW) | Label | 6FM200D | Conv. (kW) | Initial Capital | Operating Cost ($/yr) | Total NPC | COE ($/kWh) | Ren. Frac. | Diesel (L) | Label (hrs) |
|---|---|---|---|---|---|---|---|---|---|---|---|---|---|
| | | 2 | | 1 | 5 | 1 | $ 7,083 | 1,642 | $ 25,922 | 0.396 | 0.45 | 1,265 | 3,837 |
| | | 3 | 1 | 1 | 2 | 1 | $ 11,000 | 1,499 | $ 28,194 | 0.430 | 0.60 | 1,169 | 3,584 |
| | | | | 1 | 12 | 1 | $ 5,500 | 2,411 | $ 33,156 | 0.506 | 0.00 | 2,203 | 6,811 |
| | | | 1 | 1 | 8 | 1 | $ 7,500 | 2,334 | $ 34,266 | 0.523 | 0.08 | 2,025 | 6,223 |

**Fig. 5.11** Optimal hybrid renewable energy system components

The detailed cash flow for the optimum design is shown in Table 5.5. PV-diesel hybrid renewable energy system achieves the best results for cost reduction without the wind turbine. The main reason behind the result is the low average wind speed.

In addition to the cost consideration, the hybrid renewable energy system performance is examined in Figs. 5.12 and 5.13. In Fig. 5.12, the hybrid renewable energy system electrical performance is examined with total PV generation of 46% comparing with 54% of diesel generation. Also, the total excess electricity per year is 376 kWh/year which represents about 5.38% of the total generated power from the hybrid renewable energy system.

In Fig. 5.13, the solar PV module performance is illustrated. With a rated capacity of 2 kW and mean output power of 8.83 kWh/day, the total PV production is 3223 kWh/year. Also, the total hours of operation is 4470 h/year.

In Fig. 5.14, the yearly diesel generator performance is shown. With a 3756 h of operation and total generated power of 3761 kWh/year, the diesel generator contributes with 54% of the total power production of the hybrid renewable energy system. Moreover, the generator yearly fuel consumption is 1241 L with a mean electrical efficiency of 30.8%.

**Table 5.5** Project cash flow summary

|  | Capital ($) | Replace ($) | OMC ($) | Fuel ($) | Salvage ($) | Total ($) |
|---|---|---|---|---|---|---|
| PV | 3333 | 0 | 0 | 0 | 0 | 3333 |
| Diesel Generator | 1500 | 3154 | 2159 | 4983 | −367 | 11,430 |
| Vision 6FM200D | 1250 | 3923 | 1147 | 0 | −179 | 6141 |
| Convertor | 1000 | 417 | 1147 | 0 | −208 | 2356 |
| System | 7083 | 7494 | 4453 | 4983 | −754 | 23,261 |

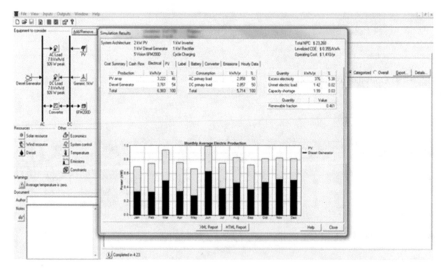

**Fig. 5.12** Hybrid renewable energy system electrical performance

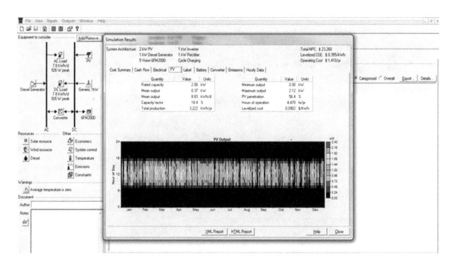

**Fig. 5.13** Solar PV module performance

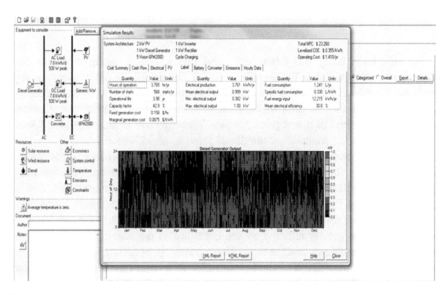

**Fig. 5.14** Diesel generator electrical performance

In Fig. 5.15, the battery storage system performance is shown with detailed analysis of battery state of charge, battery lifetime and utilized battery bank size. The battery expected lifetime is 2.69 years with energy in and energy out of 1897 and 1523 kWh/year respectively. Furthermore, the average energy cost of the battery bank (5 batteries) is estimated to be 0.062 $/kWh.

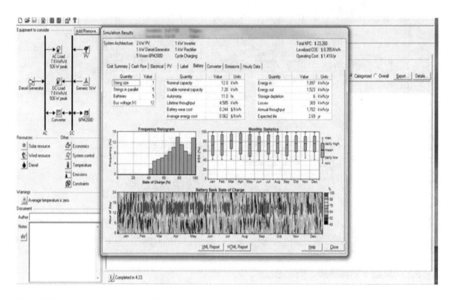

**Fig. 5.15** Battery electrical performance

An additional study has been performed in the system along with the cost objective function which is the sensitivity study. The sensitivity study measures and analyzes the effect of changing the values of system inputs on the optimum design components such as changing of the average annual wind speed, average annual solar irradiation level or diesel price per liter. The objective is to achieve the lowest cost and it is mainly focused on the diesel price as an indicator.

Hence, the sensitivity analysis is carried out on the hybrid renewable energy system by varying the fuel price (diesel) per liter and running the optimization study for all the proposed values. Also, the sensitivity analysis has another variable to be considered during the simulation which is the changing in the nominal operating temperature of the solar PV modules. As discussed in Sect. 5.2.1, the efficiency of the solar panel is greatly affected by the changing in the nominal operating temperature of the ambient environment. Hence, various operating temperatures are considered for the operation of the hybrid renewable energy system. The effects of both sensitivity factors on the performance of the hybrid renewable energy system are illustrated in Fig. 5.16.

The solar PV total electrical production versus the diesel generator electrical production is examined using a sensitivity factor and varying the diesel price as well as the nominal operating temperatures for the solar panels. It can be seen that after a diesel price of 0.5 $/L, the value of the solar production increases and the value of the diesel production decreases. In other words, the optimum design of the

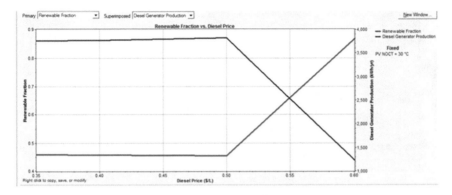

**Fig. 5.16** Renewable fraction versus diesel generator production

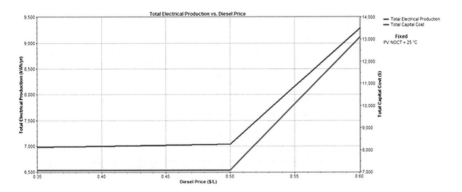

**Fig. 5.17** Total capital cost of the system versus total electrical production

hybrid renewable energy system at above 0.5 $/L for diesel price is aiming towards the renewable sources of energy which can maintain the system supply with the lowest cost.

However, by increasing the renewable energy production of the system, the capital cost increases as a result of the commissioning and installation costs of new solar panels as in Fig. 5.17. On the contrary, the annual operation and maintenance cost of the hybrid renewable energy system is significantly decreased by implementing renewable sources as indicated in Fig. 5.18.

The annual operation and maintenance costs are significantly decreased after 0.5 $/L for the diesel price because the hybrid renewable energy system implements more renewable sources approach, which decreases the fuel consumption and consequently the annual operating costs.

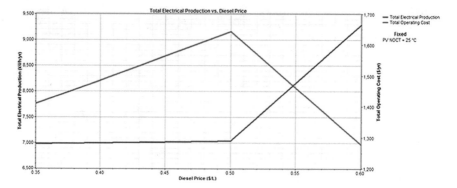

**Fig. 5.18** Annual operation and maintenance cost versus total production

## 5.7 Summary

Hybrid renewable energy system is suitable for installation and operation in the offshore environment of Malaysia. Moreover, the modeling study as well as the optimization study suggest that the hybrid renewable energy system efficiency can be significantly increased by a proper and efficient utilization of the solar PV model in the system. Such an implementation would require more efficient solar panels, a precise charging and converting control system and a well sized battery bank to preserve the excess generated power.

In addition to that, wind turbines utilization in Malaysia has proven to be less efficient than the solar PV model due to the low average wind speed in the country as compared to other parts of the world. However, by enhancing control system of the implemented wind turbines, more energy can be harvested from the wind speed and hence a consistent, cheap and powerful supply of electricity could be ensured in future.

Furthermore, the diesel generator has proven to be a main backup for the hybrid renewable energy system especially during the unfavorable climatic conditions which may affect the operation of wind turbines as well as solar PV panels. Also, the selection and sizing of the diesel generator, along with other hybrid renewable energy system components, should be carried out after a detailed study about the offshore installation current load profile taking into considerations the future loads as well as the decreasing efficiency of the system while increasing the project lifetime.

Likewise, the optimization study has concluded that the fuel price of the diesel generator set will affect greatly the operation of the hybrid renewable energy system. By increasing the fuel price of the diesel generator, the dependence on the generator in supplying the offshore installations loads will be decreased. On the contrary, the renewable fraction of the hybrid renewable energy system increases by installing new solar panels or wind turbines, which throughout the life time will be more economically justified than that by using diesel generator.

Finally, the implementation of the proposed hybrid renewable energy system in the Malaysian offshore environment would have significant effects on offshore oil and gas operations. The advantage is not only will be limited to the total reduction in costs but also reducing the emissions and preserve the environment.

# References

1. D.B. Evans, E.F. Yong, Environmental issues in oil and gas development in Malaysia, in *SPE Health, Safety and Environment in Oil and Gas Exploration and Production Conference*, Jakarta, Indonesia, 25–27 January 1994
2. C.K. Seong, T.Y. Hong, A review of offshore oil fields development in Malaysia, in *SPE Asia Pacific Oil and Gas Conference*, Kuala Lumpur, Malaysia, 20–22 March 1995
3. C.C. Wan, Use of unmanned platforms in an offshore environment, in *Offshore Technology Conference*, Houston, Texas, 2–5 May 1988
4. R. Rongsopa, Hybrid power generation for offshore wellhead platform: a starting point for offshore green energy, in *International Petroleum Technology Conference*, Bangkok, Thailand, 15–17 November 2011
5. W. Zhou, Simulation and optimum design of hybrid solar-wind and solar-wind-diesel power generation systems, Ph.D. dissertation (Hong Kong Polytechnic University), 2008
6. G. Walker, Evaluating MPPT converter topologies using a MATLAB PV model. J. Electr. Electron. Eng., 49–56 (2001) (Australia)
7. W. Schmitt, Modeling and simulation of photovoltaic hybrid energy systems optimization of sizing and control, in *Photovoltaic Specialists Conference* (New Orleans, LA, USA, 19–24 May 2002), pp. 1656–1659
8. S. M. S., Modeling and simulation of photovoltaic module using MATLAB/Simulink. Int. J. Chem. Environ. Eng., **2** (2011)
9. R.H. Harrison, E. Snel, Large wind turbines (Wiley, England, 2000)
10. A. D. G. a. U. Hassan, What is power quality?, in *Proceedings of the 10-th British Wind Energy Association Conference*, ed. by D.J. Milborrow, London, March 1998, pp. 22–24
11. R.N.B.S.A. Jalilvand, Design of a multilevel control strategy for integration of stand-alone wind/diesel system. Int. J. Electr. Power Energy Syst. **35**, 123–137 (2012)
12. K. Sopian, A.H. Haris, D. Rouss, M.A. Yusof, Building Integrated Photovoltaic (BIPV) in Malaysia—Potential, current status strategies for long term cost reduction. ISESCO J. Sci. Technol. **1** (2005)
13. M.Z. Hussin, A. Yaacob, R.A. Rahman, Z.M. Zain, S. Shaari, A.M. Omar, Monitoring results of Malaysian building integrated PV project in grid-connected photovoltaic system in Malaysia. Energy Power **2**, 39–45 (2012)
14. M.A. Elhadidy, S.M. Shaahid, Parametric study of hybrid (wind + solar + diesel) power generating systems. Renew Energy **21**, 129–139 (2000)
15. F. Sulaiman, Renewable energy and its future in Malaysia: a country paper, in *Proceedings of Asia-Pacific Solar Experts Meeting*, 1995
16. E.P. Chiang, Z.A. Zainal, P.A. Aswatha Narayana, K.N. Seetharamu, The potential of wave and offshore wind energy in around the coastline of Malaysia that face the South China Sea, in *Proceedings of The International Symposium on Renewable Energy: Environment Protection & Energy Solution for Sustainable Development*, Kuala Lumpur, Malaysia, 14–17 September 2003
17. M. Kalogera, P. Bauer, Optimization of an off-grid hybrid system for supplying offshore platforms in Arctic climates, in *The 2014 International Power Electronics Conference*, Hiroshima, Japan, 18–21 May 2014

# Chapter 6
# Effect of Simpler Neill's Mapping Function Models Using Non Linear Regression Method on Tropospheric Delay

**Hamzah Sakidin, Siti Rahimah Batcha and Asmala Ahmad**

## 6.1 Introduction

The calculation of tropospheric delay is in distance, and a regular tropospheric delay would be 2.5 m. This defines that the troposphere makes a GPS range observation obtain clear additional 2.5-m distance among the ground based receiver and a satellite at zenith [1].

The Tropospheric Delay (TD) at any elevation angles can be separated in terms of the zenith delays and mapping functions. This form allows the use of different mapping functions for the hydrostatic and non-hydrostatic delay components. The mapping functions depend on the elevation angles, whereby at 90° of elevation angle, the scale factor of the mapping function value is 1. This value generates small number for the tropospheric delay (TD) [2], which is defined by:

$$TD = ZHD.m_h(\varepsilon) + ZWD.m_w(\varepsilon) \tag{6.1}$$

where ZHD is zenith hydrostatic delay $(m_h)$, ZWD is zenith wet delay $(m_w)$, $m_h(\varepsilon)$ is hydrostatic mapping function, and $m_w(\varepsilon)$ is wet mapping function.

The definition of mapping function stated as the signal delay that go through the neutral atmosphere [3]. The most popular parameters in mapping function are temperature, pressure and relative humidity. The established mapping function models to calculate tropospheric delay values are given in a form of continued fractions. Most of the modern mapping function models have separate mapping

H. Sakidin (✉) · S. R. Batcha
Department of Fundamental and Applied Sciences,
Universiti Teknologi PETRONAS, Seri Iskandar, Malaysia
e-mail: hamzah.sakidin@utp.edu.my

A. Ahmad
Faculty of Information Technology & Communication,
Universiti Teknikal Malaysia Melaka, Durian Tunggal, Melaka, Malaysia

S. A. Sulaiman et al. (eds.), *Sustainable Electrical Power Resources through Energy Optimization and Future Engineering*, SpringerBriefs in Energy,
https://doi.org/10.1007/978-981-13-0435-4_6

functions for the hydrostatics (dry) and the non-hydrostatics (wet) components. Hydrostatic and wet parts of the mapping function are dissimilar due to thickness of the troposphere [4]. To calculate the delay due to the troposphere and to get a high degree of precision, mapping function is needed in the evaluation process [5].

## 6.2   Neill's Mapping Function

The most frequent used mapping function is Neill that was established in 1996. It was strongly believed to be the most precise at elevation angle lower than 10 and does not need meteorological observations [6]. Neill [7], kept the basic form of the Herring (MTT model—Herring, 1992) mapping function adding a height correction term and assuming that the elevation dependence is a function of only geographical parameters (if we accept that in a way the day of the year is also a constant and independent parameter) and proposed the function.

For Neill's mapping function (hydrostatic component) there are 26 operations in the model, which is given by:

$$NMF_h(\varepsilon) = \frac{1 + \frac{a}{1 + \frac{b}{1+c}}}{\sin \varepsilon + \frac{a}{\sin \varepsilon + \frac{b}{\sin \varepsilon + c}}} + \left[ \frac{1}{\sin \varepsilon} - \left( \frac{1 + \frac{a_{ht}}{1 + \frac{b_{ht}}{c_{ht}}}}{\sin \varepsilon + \frac{a_{ht}}{\sin \varepsilon + \frac{b_{ht}}{\sin \varepsilon + c_{ht}}}} \right) \right] H \qquad (6.2)$$

where $\varepsilon$ is elevation angle (radian), and $H$ is the station height above sea level (km). For Neill's mapping function (wet component) there are 11 operations which are addressed by:

$$NMF_w(\varepsilon) = \frac{1 + \frac{a_w}{1 + \frac{b_w}{1+c_w}}}{\sin \varepsilon + \frac{a_w}{\sin \varepsilon + \frac{b_w}{\sin \varepsilon + c_w}}} \qquad (6.3)$$

## 6.2.1   Simplification of Hydrostatics Neill Mapping Function

The original hydrostatics Neill's mapping function model has been named as A, stated in Eq. (6.2). This model is simplified and is named as A1:

$$A1 = K + \frac{L}{\varepsilon} \qquad (6.4)$$

where $\varepsilon$ is the elevation angle (degree), constants $K$ is 0.583 and $L$ is 37.491.

Table 6.1 shows the data for the original hydrostatics Neill's mapping function and also the data for the simpler hydrostatics Neill's mapping function. The

| E | A | A1 | $X = \frac{A - A1}{A} \times 100\%$ | $|X|$ |
|---|---|---|---|---|
| 3 | 18.58 | 19.33 | −4.04 | 4.04 |
| 5 | 10.15 | 8.08 | 20.39 | 20.39 |
| 10 | 5.56 | 4.33 | 22.12 | 22.12 |
| 15 | 3.8 | 3.08 | 18.95 | 18.95 |
| 20 | 2.9 | 2.46 | 15.17 | 15.17 |
| 25 | 2.35 | 2.08 | 11.49 | 11.49 |
| 30 | 1.99 | 1.83 | 8.04 | 8.04 |
| 35 | 1.74 | 1.65 | 5.17 | 5.17 |
| 40 | 1.55 | 1.52 | 1.94 | 1.94 |
| 45 | 1.41 | 1.42 | −0.71 | 0.71 |
| 50 | 1.3 | 1.33 | −2.31 | 2.31 |
| 55 | 1.22 | 1.27 | −4.10 | 4.1 |
| 60 | 1.15 | 1.21 | −5.22 | 5.22 |
| 65 | 1.1 | 1.16 | −5.45 | 5.45 |
| 70 | 1.06 | 1.12 | −5.66 | 5.66 |
| 75 | 1.04 | 1.08 | −3.85 | 3.85 |
| 80 | 1.02 | 1.05 | −2.94 | 2.94 |
| 85 | 1.00 | 1.02 | −2.00 | 2 |
| 90 | 1.00 | 1.00 | 0.00 | 0 |
| | | | Average $|X|$ | 7.34 |

**Table 6.1** Comparison data for original and simpler Neill's mapping function (hydrostatics component)

calculation of the absolute relative error| and also its percentage between the original and the simpler model can give the comparison between the two models.

The average of percentage of absolute relative error between the original and simpler Neill's (hydrostatics) mapping function models is 7.34% as given in the Table 6.1. It shows that the different between the original model and the simpler model for hydrostatics component is very small.

## 6.2.2  Simplification of Wet Neill's Mapping Function

The original wet Neill's mapping function model is named as B, as stated in Eq. (6.3). This model is simplified and has been named as B1 as shown in Eq. (6.5). For wet Neill's mapping function, the model is not as complicated as the model in hydrostatic component. From 3° to 90° elevation angle, the simpler model named as B1, which reduced 11 operations to only two operations, is given by:

$$B1 = M + \frac{N}{\varepsilon} \qquad (6.5)$$

where $\varepsilon$ is elevation angle (degree), constants $M$ is 0.513 and $N$ is 43.859.

**Table 6.2** Comparison data for original and simpler Neill's mapping function (wet component)

| E | B | B1 | $Y = \frac{B-B1}{B} \times 100\%$ | $|Y|$ |
|---|---|---|---|---|
| 3 | 21.854 | 22.443 | −2.70 | 2.7 |
| 5 | 10.751 | 9.285 | 13.64 | 13.64 |
| 10 | 5.657 | 4.899 | 13.40 | 13.4 |
| 15 | 3.833 | 3.437 | 10.33 | 10.33 |
| 20 | 2.911 | 2.706 | 7.04 | 7.04 |
| 25 | 2.36 | 2.267 | 3.94 | 3.94 |
| 30 | 1.997 | 1.975 | 1.10 | 1.1 |
| 35 | 1.741 | 1.766 | −1.44 | 1.44 |
| 40 | 1.554 | 1.609 | −3.54 | 3.54 |
| 45 | 1.413 | 1.488 | −5.31 | 5.31 |
| 50 | 1.305 | 1.39 | −6.51 | 6.51 |
| 55 | 1.22 | 1.31 | −7.38 | 7.38 |
| 60 | 1.554 | 1.244 | 19.95 | 19.95 |
| 65 | 1.103 | 1.188 | −7.71 | 7.71 |
| 70 | 1.064 | 1.14 | −7.14 | 7.14 |
| 75 | 1.035 | 1.098 | −6.09 | 6.09 |
| 80 | 1.015 | 1.061 | −4.53 | 4.53 |
| 85 | 1.004 | 1.029 | −2.49 | 2.49 |
| 90 | 1 | 1 | 0.00 | 0 |
|  |  |  | Average $|Y|$ | 6.54 |

Table 6.2 shows the data for the original wet Neill's mapping function and also the data for the simpler wet Neill's mapping function. The calculation of the absolute relative error and also its percentage between the original and the simpler model can give the comparison between the two models.

The average of percentage of absolute relative error between the original and simpler Neill's (wet) mapping function models is 6.54% as given in the Table 6.2. It shows that the different between the original model and the simpler model for hydrostatics component is very small.

## 6.3 Tropospheric Delay for Simpler Neill Mapping Function Models

Table 6.3 shows the result of tropospheric delay for simpler Neill's mapping function by using Eq. (6.1). The value for A1 (hydrostatics component) need to be multiplied by 2.3 m and the value for B1 (wet component) need to be multiplied by 0.25 m [8].

**Table 6.3** Tropospheric delay using simpler Neill mapping function

| Elevation angle, $\varepsilon$ (degree) | A1 × 2.3 m | B1 × 0.25 m | Total tropospheric delay (m) |
|---|---|---|---|
| 3 | 44.456 | 5.611 | 50.067 |
| 5 | 18.587 | 2.321 | 20.908 |
| 10 | 9.964 | 1.225 | 11.189 |
| 15 | 7.090 | 0.859 | 7.949 |
| 20 | 5.652 | 0.676 | 6.328 |
| 25 | 4.790 | 0.567 | 5.357 |
| 30 | 4.215 | 0.494 | 4.709 |
| 35 | 3.805 | 0.442 | 4.247 |
| 40 | 3.497 | 0.402 | 3.899 |
| 45 | 3.257 | 0.372 | 3.629 |
| 50 | 3.065 | 0.348 | 3.413 |
| 55 | 2.909 | 0.328 | 3.237 |
| 60 | 2.778 | 0.311 | 3.089 |
| 65 | 2.668 | 0.297 | 2.965 |
| 70 | 2.573 | 0.285 | 2.858 |
| 75 | 2.491 | 0.274 | 2.765 |
| 80 | 2.419 | 0.265 | 2.684 |
| 85 | 2.355 | 0.257 | 2.612 |
| 90 | 2.300 | 0.250 | 2.550 |

Table 6.4 shows the calculation of tropospheric delay for using both original and simpler Neill's mapping function. The relative error of tropospheric delay values by using original and simpler Neill's mapping function values are given by:

$$\text{Relative Error} = \frac{153.85 - 144.455}{153.85} = 0.061 \tag{6.6}$$

Hence, the relative error percentage is 6.1%.

Figure 6.1 shows the graph of tropospheric delay using original Neill and the simpler Neill model. It is shown the figure that there is no significant difference between the two graphs.

## 6.4 Summary

The average percentage of absolute relative error between the original and simpler Neill's (hydrostatics) mapping function models is 7.34%. It shows that the difference between the original model and the simpler model for hydrostatics component is very small.

**Table 6.4** Tropospheric delay for original Neill's and simpler Neill mapping function

| Elevation angle, $\varepsilon$ (degree) | Tropospheric delay for original Neill's | Tropospheric delay for simpler Neill's |
|---|---|---|
| 3 | 48.200 | 50.067 |
| 5 | 26.035 | 20.908 |
| 10 | 14.193 | 11.189 |
| 15 | 9.703 | 7.949 |
| 20 | 7.393 | 6.328 |
| 25 | 6.002 | 5.357 |
| 30 | 5.083 | 4.709 |
| 35 | 4.435 | 4.247 |
| 40 | 3.960 | 3.899 |
| 45 | 3.603 | 3.629 |
| 50 | 3.325 | 3.413 |
| 55 | 3.111 | 3.237 |
| 60 | 2.943 | 3.089 |
| 65 | 2.813 | 2.965 |
| 70 | 2.713 | 2.858 |
| 75 | 2.639 | 2.765 |
| 80 | 2.588 | 2.684 |
| 85 | 2.560 | 2.612 |
| 90 | 2.550 | 2.550 |
| Total | 153.85 | 144.455 |

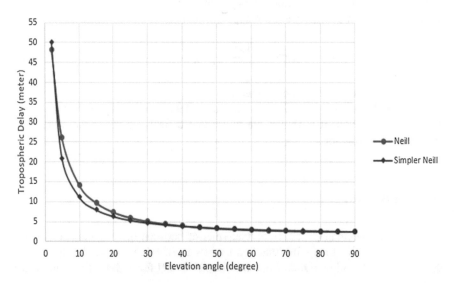

**Fig. 6.1** Comparison between total tropospheric delay using original Neill and simpler Neill mapping function model

The average percentage of absolute relative error between the original and simpler Neill's (wet) mapping function models is 6.54%. It shows that the difference between the original model and the simpler model for hydrostatics component also is very small and not significant.

The tropospheric delay values for original Neill's mapping function model is 153.85 m and for simpler Neill's model is 144.46 m. It shows the comparison relative error between the two models is only 6.1% and shows no significant difference with the original Neill's mapping function. The number of operation for the Neill's hydrostatics component model can be reduced from 26 operations to 2 operations. However, the number of operations for Neill's wet component model can be reduced from 11 operations to 2 operations. As a conclusion, the simpler Neill's mapping function models can represent the original Neill's mapping function due to no significant effect on the tropospheric delay value.

# References

1. S. Nistor, A.S. Buda, Determination of zenith tropospheric delay and precipitable water vapor using GPS technology. Math. Model. Civil Eng. **12**(1), 21–26 (2016)
2. A. Farah, for near-equatorial-tropospheric delay correction **5**(2), 67–72 (2011)
3. M. Jgouta, B. Nsiri, R. Marrakh, C. Author, Usage of a correction model to enhance the evaluation of the zenith tropospheric delay. Int. J. Appl. Eng. Res. **11**(6), 4648–4654 (2016)
4. S. Younes, A. Elmezayen, Science A comprehensive comparison of atmospheric mapping functions for GPS measurements in abstract. Geod. Sci. **2**(3), 216–223 (2012)
5. M. Elsobeiey, M. El-diasty, Impact of tropospheric delay gradients on total tropospheric delay and precise point positioning. Int. J. Geosci. **7**, 645–654 (2016)
6. C. Rocken, S. Sokolovskiy, J.M. Johnson, D. Hunt, Improved mapping of tropospheric delays, 1205–1213 (2001)
7. A.E. Niell, Global mapping functions for the atmosphere delay at radio wavelengths. J. Geophys. Res. **101**, 3227 (1996)
8. L. Combrinck, A review of the lunar laser ranging technique and contribution of timing systems. **112**(3), 1–9 (2016)

Printed in the United States
By Bookmasters